Lecture Notes in Mathematics

Edited by A. Dold and B. Eckmann

1045

Differential Geometry

Proceedings of the International Symposium
held at Peñíscola, Spain, October 3–10, 1982

Edited by A. M. Naveira

Springer-Verlag
Berlin Heidelberg New York Tokyo 1984

Editor

Antonio M. Naveira
Departamento de Geometría y Topología Facultad de Matemáticas
Burjasot, Valencia, Spain

AMS Subject Classifications (1980): 22E, 53A, 53B, 53C, 57R, 57S, 58A, 58B, 58C, 58G, 81E

ISBN 3-540-12882-4 Springer-Verlag Berlin Heidelberg New York Tokyo
ISBN 0-387-12882-4 Springer-Verlag New York Heidelberg Berlin Tokyo

Library of Congress Cataloging in Publication Data. Main entry under title: Differential geometry. (Lecture notes in mathematics; 1045) In English and French.
Proceedings of the Symposium on Differential Geometry. 1. Geometry, Differential–Congresses. I. Naveira, A.M. (Antonio Martínez), 1940-. II. Symposium on Differential Geometry (1982: Peñíscola, Spain) III. Series: Lecture notes in mathematics (Springer-Verlag); 1045.
QA3.L28 no. 1045 [QA641] 510s [516.3'6] 83-20457
ISBN 0-387-12882-4 (U.S.)

Printing and binding: Beltz Offsetdruck, Hemsbach/Bergstr.
2146/3140-543210

PREFACE

The present volume includes the texts of all lectures given at a Symposium on Differential Geometry which was held at Peñíscola, Spain, from October 3 to 10, 1982. The Symposium was attended by some forty mathematicians from all over the world.

There have been five International Symposia on Differential Geometry in Spain during the last twenty years. One of them took place in Salamanca (1979), and the remaining four in Santiago de Compostela (1962, 1967,1972,1978). Of those, the first three ones were organized by Prof. E. Vidal Abascal and the last one was held in his hommage at his retirement. Our wish when organizing this Symposium has been to continue with this tradition initiated by Prof. Vidal Abascal to whom we express our deepest gratitude for his exemplary contribution to the development of Differential Geometry in Spain. We hope to be able to continue with this tradition by holding these meetings periodically.

The Organizing Commitee of this Symposium is glad to express its gratitude to all those who contributed to the success of the meeting and in particular to all participants. We also wish to express our sincere thanks to the Ministerio de Educación, to the Facultad de Matemáticas de Valencia, to the Diputación Provincial de Castellón for their generous financial support, and to the Instituto de Administración Local de Peñíscola for allowing us to use their premises during the Symposium. The facilities given by Springer-Verlag, and the co-operation of Prof. S. I. Andersson in our early contacts with the publishers are also gratefully acknowledged.

Finally we wish to thank F. Marhuenda and J. Monterde for their careful typing of most of the manuscripts.

A. M. Naveira

President of the Organizing Commitee

LIST OF PARTICIPANTS

E. Abbena
U. Torino, Italy

S. I. Andersson
U. Clausthal, West Germany

M. Asorey
U. Zaragoza, Spain

M. A. Baratta
U. Parma, Italy

M. Bendala
U. Sevilla, Spain

D. Bernard
U. Strasbourg, France

J. L. Cabrerizo
U. Sevilla, Spain

F. Carreras
U. Valencia, Spain

D. Chinea
U. La Laguna, Spain

L. A. Cordero
U. Santiago, Spain

C. Currás-Bosch
U. Barcelona, Spain

S. Donnini
U. Parma, Italy

F. J. Echarte
U. Sevilla, Spain

J. Etayo
U. Madrid, Spain

M. Fernández
U. Santiago, Spain

M. Fernández-Andrés
U. Sevilla, Spain

A. Ferrández
U. Valencia, Spain

E. Fossas
U. Barcelona, Spain

S. Garbiero
U. Torino, Italy

O. Gil Medrano
U. Valencia, Spain

J. Girbau
U. Aut. Barcelona, Spain

A. Gray
U. Maryland, U.S.A.

J. Grifone
U. Toulouse, France

R. Langevin
U. Dijon, France

A. Lichnerowicz
Collège de France, France

F. Mascaró
U. Valencia, Spain

V. Miquel
U. Valencia, Spain

A. Montesdeoca
U. La Laguna, Spain

A. Montesinos
U. Valencia, Spain

A. M. Naveira
U. Valencia, Spain

M. Nicolau
U. Aut. Barcelona, Spain

G. B. Rizza
U. Parma, Italy

A. H. Rocamora
U. Pol. Valencia, Spain

C. Romero
U. Valencia, Spain, and
Southampton, United Kingdom

A. Ros
U. Granada, Spain

M. Sekizawa
U. Tokyo, Japan

R. Sivera
U. Valencia, Spain

F. Torres-Lopera
U. Santiago, Spain

L. Vanhecke
U. Leuven, Belgium

F. Varela
U. Murcia, Spain

J. L. Viviente
U. Zaragoza, Spain.

TABLE OF CONTENTS

PSEUDODIFFERENTIAL OPERATORS AND CHARACTERISTIC CLASSES FOR NON-ABELIAN COHOMOLOGY

Stig I. Andersson

Institut für Theoretische Physik
der Technischen Univ. Clausthal
D-3392 Clausthal-Zellerfeld, FRG.

0. Introduction

The object of study in this work is the interplay between analytic properties of pseudodifferential operators (psdo) on vector bundles and the geometry of the vector bundles themselves.

Let X be a connected smooth n-manifold and $E_1 \xrightarrow[p_1]{} X$, $E_2 \xrightarrow[p_2]{} X$ and $E \xrightarrow[p]{} X$ vector bundles of dimensions m_1, m_2 and m respectively. By $S(E_1)$, $S(E_2)$ and $S(E)$ we denote the smooth sections and $PDiff_k(E_1,E_2)$ ($Diff_k(E_1,E_2)$) stands for the psdo (partial differential operators, pdo) of order k, mapping $S(E_1) \longrightarrow S(E_2)$.

Given $P \in PDiff_k(E_1,E_2)$, by a lifting process we shall attach a <u>connection operator</u> $Q: S(E_1 \otimes E) \longrightarrow S(E_2 \otimes E)$. Furthermore, by a procedure analogous to the Bott-Chern-Weil construction, we shall develop a naturally associated theory of characteristic classes, based however on a <u>non-abelian cohomology theory</u>.

Modulo problems in the relevant homological algebra, forcing us to consider only the one and two dimensional cohomology sets, this provides a bridge between analytic properties of P and the vector bundle geometry. The construction extends the one of Asada's (AS 1) for local operators.

1. Asada Connections for $\mathrm{Diff}_k(E_1,E_2)$

To provide motivation for our later construction, we shall briefly review the essential steps in the construction of an Asada connection to a given pdo.

$P \in \mathrm{Diff}_k(E_1,E_2)$ being a local object, taking restrictions to open sets is a transitive operation, hence the commutative diagram (U open set)

$$\begin{array}{ccc} S(E_1) & \xrightarrow{\;P\;} & S(E_2) \\ \downarrow & & \downarrow \\ S(E_1 \restriction U) & \xrightarrow[P\restriction U]{} & S(E_2 \restriction U) \end{array}$$

Let (U_i,μ_i) be a locally trivializing atlas for E_1 and E_2 on X and write $P_i := P\restriction U_i$. A local trivialization for $E_1 \xrightarrow[p_1]{} X$ is then a VB-equivalence

$$\begin{array}{ccc} p_1^{-1}(U_i) & \xrightarrow{\;F_i^1\;} & \mu_i(U_i) \times \mathbb{C}^{m_1} \\ {\scriptstyle p_1}\downarrow & & \downarrow{\scriptstyle \text{proj.}} \\ U_i & \xrightarrow[\mu_i]{} & \mu_i(U_i) \end{array}$$

where $F_i^1 : p_1^{-1}(U_i) \longrightarrow U_i \times C^{m_1} \xrightarrow[(\mu_i \times \mathrm{Id})]{} \mu_i(U_i) \times \mathbb{C}^{m_1}$ defines the fiber chart.

For the sections we obtain the induced isomorphisms

$$Z_i^1 : S(p_1^{-1}(U_i)) \longrightarrow [S(\mu_i(U_i))]^{m_1} \;;\; Z_i^2 : S(p_2^{-1}(U_i)) \longrightarrow [S(\mu_i(U_i))]^{m_2}$$

(and analogously for S_c , the sections with compact support). So locally P induces the operator $P_i' : [S(U_i')]^{m_1} \longrightarrow [S(U_i')]^{m_2}$, $U_i' := \mu_i(U_i) \subset R^n$, by

$$[S(U_i')]^{m_1} \xrightarrow[(Z_i^1)^{-1}]{} S(p_1^{-1}(U_i)) \xrightarrow[P_i]{} S(p_2^{-1}(U_i)) \xrightarrow[Z_i^2]{} [S(U_i')]^{m_2}.$$

For two charts U_i and U_j with $U_{ij} := U_i \cap U_j \neq \emptyset$, we thus obtain the following commutative diagram, describing the compatibility condition on U_{ij};

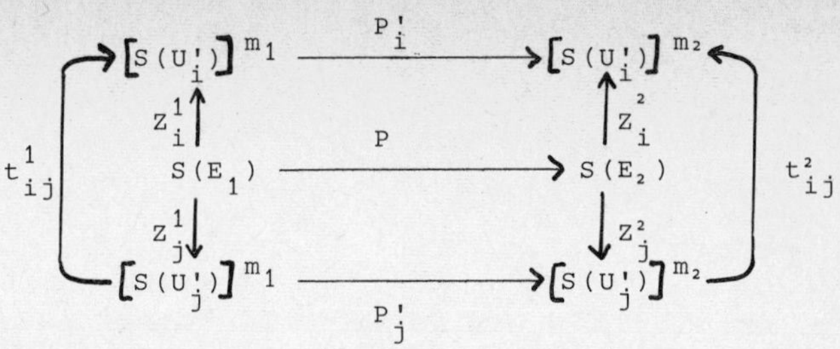

with the transition functions $t^1_{ij} := Z^1_i \circ (Z^1_j)^{-1}$, $t^2_{ij} := Z^2_i \circ (Z^2_j)^{-1}$ for E_1 and E_2 respectively.

Define the (index-preserving) lifting $\tilde{P}$ of P by $\tilde{P}_i : S(E_1 \otimes E \restriction U_i) \longrightarrow S(E_2 \otimes E \restriction U_i)$ such that $\sigma_k(\tilde{P}) = \sigma_k(P) \otimes \mathrm{Id}_E$ ($\sigma_k(P)$ = symbol of P etc). Explicitly in a chart, writing $P'_i = \sum_{|\alpha| \leq k} a_{\alpha,i} D^{\alpha}_i$ (α a multi-index) we define the lifting to be

$$\tilde{P}'_i := \sum_{|\alpha| \leq k} (a_{\alpha,i} \otimes \mathrm{Id}_E)\, D^{\alpha}_i .$$

In general the compatibility requirements are violated though for the lifting, i.e.

$$\tilde{P}'_i \, T^1_{ij} \neq T^2_{ij} \, \tilde{P}'_j \quad \text{on } U_{ij}$$

where $T^1_{ij} := t^1_{ij} \otimes e_{ij}$, $T^2_{ij} := t^2_{ij} \otimes e_{ij}$, ($e_{ij}$ = transition function for E) are the transition functions for $E_1 \otimes E$ and $E_2 \otimes E$ respectively.

<u>Definition</u>: $Q = \{Q_i\}$ with $Q_i \in \mathrm{Diff}_r(E_1 \otimes E \restriction U_i, E_2 \otimes E \restriction U_i)$ for $r < k$, is an <u>Asada E-connection of P</u> iff $(\tilde{P} + Q) \in \mathrm{Diff}_k(E_1 \otimes E, E_2 \otimes E)$ i.e.

$$(\tilde{P}'_i + Q'_i) \circ T^1_{ij} = T^2_{ij} \circ (\tilde{P}'_j + Q'_j) \quad \text{on } U_{ij}.$$

The obstruction $W_{ij} := Q'_i T^1_{ij} - T^2_{ij} Q'_j$ can be simply computed. Putting $Q'_i = \sum_{|\alpha| \leq k-1} q_{\alpha,i} D^{\alpha}_i$, we obtain $W_{ij} = \sum_{\beta} e^{\alpha}_{\beta} D^{\beta}_i$ where

$$e^{\alpha}_{\beta} = \sum_{\substack{\alpha \\ |\beta| \leq |\alpha| \leq k}} \binom{\alpha}{\beta} (a_{\alpha,i} \otimes \mathrm{Id}_E) \left\{ D^{\alpha-\beta}_i t^1_{ij} \otimes e_{ij} - D^{\alpha-\beta}_i (t^1_{ij} \otimes e_{ij}) \right\} =$$

$$= \sum_{|\beta| \leq |\alpha| \leq k-1} \binom{\alpha}{\beta} q_{\alpha,i} D^{\alpha-\beta}_i T^1_{ij} - T^2_{ij} \, q_{\beta,j} \, (\mathrm{Jac}_{ij})^{\beta},$$ with Jac_{ij} = the Jacobian for the change of coordinates.

On the symbol level, $\sigma_{k-1}(W_{ij}) = \sum_{\beta} e^{\alpha}_{\beta} \xi^{\beta}_i$.

Assuming $0 = \sigma_{k-1}(W) = \ldots\ldots = \sigma_{k-j}(W)$; there exists an E-connection of P of order $\leq k-(j+2)$ iff $\sigma_{k-(j+1)}(W) = 0$.

Here $\sigma_{k-j}(W) = \{\sigma_{k-j}(W_{ij})\}$ and we call $\sigma_{k-(j+1)}(W)$ the obstruction of order $k-(j+2)$.

Computing, we obtain explicitly; assuming $0 = \sigma_{k-1}(W_{ij}) = \ldots = \sigma_{k-s+1}(W_{ij})$

$$\sigma_{k-s}(W_{ij}) = -\sum_{|\beta|=k-s}\left[\sum_{(k-s)\leq|\alpha|\leq k}\binom{\alpha}{\beta}(a_{\alpha,i}\,\xi_i^{\alpha}\otimes Id_E)\sum_{|\gamma|\geq 1}\binom{\alpha-\beta}{\gamma}\left[D_i^{\alpha-\beta-\gamma}t_{ij}^{1}\otimes D_i^{\gamma}\,e_{ij}\right]\right]$$

$:= -M_{ij}^{s}(e_{ij})$, defining the order s differential operator M_{ij}^{s} locally.

<u>Example</u>: $s=1$, $\sigma_{k-1}(W_{ij}) = -\sum_{|\beta|=k-1}\left[\sum_{l=1}^{n}(a_{\beta+1_l,i}\,\xi_i^{\beta}\otimes\frac{\partial}{\partial x(i)_l})\right]T_{ij}^{1}$

$:= -M_i^{1}(e_{ij})(t_{ij}^{1}\otimes Id_E)$.

The Asada construction is now completed by the following series of observations;

1^0. $\sigma_{k-s}(W_{ij})$ is a smooth section in the bundle $\mathcal{G}_{k-s}\restriction U_{ij}$ where

$$\mathcal{G}_{k-s} := Hom(E_1,E_2)\otimes S^{k-s}(T^*(X))\otimes Hom(E,E).$$

2^0. $\sigma_{k-s}(W_{ij})$ is independent of Q for order$(Q)\leq k-s$.

3^0. $\sigma_{k-s}(W) = 0$ in $S(\mathcal{G}_{k-s}) \Leftrightarrow$ there exists an order $k-(s+1)$ E-connection of P.

4^0. $\sigma_{k-s}(W_{ij})$ is a special 1-cocycle (defines a Mittag-Leffler or Cousin I-distribution) since

$$\sigma_{k-s}(W_{ij})T_{jr}^{1} + T_{ij}^{2}\sigma_{k-s}(W_{jr}) = \sigma_{k-s}(W_{ir}) \text{ on } U_{ijr} := U_i\ U_j\ U_r,$$

and

$$T_{ji}^{2}\sigma_{k-s}(W_{ij})\ T_{ji}^{1} = -\sigma_{k-s}(W_{ji}) \text{ on } U_{ij}.$$

It is now natural to view $M^s = \{M_{ij}^{s}(\cdot)\}$ as a differential operator (of order s) on the sheaf $S(Hom(E,E))$ with the image sheaf $Ran(M^s) \subset S(X,\mathcal{G}_{k-s})$. In particular, $\sigma_{k-s}(W_{ij}) \in Ran(M_{ij}^{s})$.

Being a 1-cocycle, $\sigma_{k-s}(W_{ij})$ determines an <u>obstruction class</u>,

$\Sigma_{k-s} := \{[\sigma_{k-s}(W_{ij})]\}$ in $H^1(X,\mathrm{Ran}(M^s))$, i.e. we can formulate the result as follows; there exists an E-connection of P of order k-(s+1) iff $\Sigma_{k-s} = 0$ in $H^1(X,\mathrm{Ran}(M^s))$.

This means in turn that $\sigma_{k-s}(W_{ij})$ defines a 1-coboundary, so the Cousin I-problem is solvable, i.e. there exists $f = \{f_i\} \in S(\mathrm{Hom}(E,E))$ so that

$$\sigma_{k-s}(W_{ij}) = M^s_{ij}(f_i)\, T^1_{ij} - T^2_{ij}\, M^s_{ji}(f_j).$$

We have trivially the exact sequence of sheaves;

$$0 \longrightarrow \mathrm{Ker}(M^s) \longrightarrow S(X,\mathrm{Hom}(E,E)) \xrightarrow{M^s} \mathrm{Ran}(M^s) \longrightarrow 0$$

generating (less trivially) the exact cohomology sequence;

$$0 \longrightarrow H^0(X,\mathrm{Ker}(M^s)) \longrightarrow H^0(X,S(X,\mathrm{Hom}(E,E))) \longrightarrow H^0(X,\mathrm{Ran}(M^s)) \longrightarrow$$
$$\longrightarrow H^1(X,\mathrm{Ker}(M^s)) \longrightarrow H^1(X,S(X,\mathrm{Hom}(E,E))) \longrightarrow H^1(X,\mathrm{Ran}(M^s)).$$

Following Dedecker (DE), we let A^s be the elements in $\mathrm{Aut}(S(X,\mathrm{Hom}(E,E)))$ leaving $\mathrm{Ker}(M^s)$ invariant (equivalently, those automorphisms of $\mathrm{Ker}(M^s)$ which extends to automorphisms of $S(X,\mathrm{Hom}(E,E))$. We can then define a a map into a 2-cohomology set;

$$\#^s : H^1(X,\mathrm{Ran}(M^s)) \longrightarrow H^2(X,A^s)$$

and $\#^s(\Sigma_{k-s}) \in H^2(X,A^s)$ corresponds to the first Chern class.

2. Extension to Complexes in Diff.

Given a differential complex

$$(P):\quad S(E_1) \xrightarrow{P_1} S(E_2) \xrightarrow{P_2} \dots\dots\dots$$

with $P_r \in \mathrm{Diff}_k(E_r,E_{r+1})$. For a vector bundle $E \xrightarrow[p]{} X$, we define the lifted complex

$$(\tilde{P}):\quad S(E_1 \otimes E) \xrightarrow{\tilde{P}_1} S(E_2 \otimes E) \xrightarrow{\tilde{P}_2} \dots\dots\dots$$

$(\tilde{P})$ is still a complex since $\widetilde{P_{r+1} \circ P_r} = \tilde{P}_{r+1} \circ \tilde{P}_r$ but, as we have discussed in the previous section, $\tilde{P}_r$ is in general no longer a pdo. The connection operators Q_r are exactly designed to measure this obstruction. We form the sequence

$$(\tilde{P}_Q):\ S(E_1\otimes E)\xrightarrow{\tilde{P}_1+Q_1} S(E_2\otimes E)\xrightarrow{\tilde{P}_2+Q_2}\ \ldots\ldots.$$

where, by construction, $\tilde{P}_r+Q_r \in \mathrm{Diff}_k(E_r\otimes E, E_{r+1}\otimes E)$ but on the other hand, $(\tilde{P}_Q)$ is no longer a complex. The deviation in the r:th place is measured by $(\tilde{P}_{r+1}+Q_{r+1})\circ(\tilde{P}_r+Q_r) = \tilde{P}_{r+1}\circ Q_r + Q_{r+1}\circ\tilde{P}_r + Q_{r+1}\circ Q_r$ (since (P) is a complex and $\tilde{P}_{r+1}\circ\tilde{P}_r = \widetilde{P_{r+1}\circ P_r}$). This motivates, given $P\in \mathrm{Diff}_k(E_1,E_2)$ and Q, an E-connection of P, the following

Definition: The E-curvature of Q is

$$\Theta_P(Q) := \tilde{P}\circ Q + Q\circ\tilde{P} + Q^2 .$$

It is easy to see, that if (P) is a complex and Q a flat connection then $\Theta_{P_r}(Q_r) = 0$ for all r and so $(\tilde{P}_Q)$ is a complex of differential operators.

3. The Characteristic Classes for $\mathrm{PDiff}_k(E_1,E_2)$

Taking into account the pseudolocal nature of psdo, the local picture given for pdo above makes sense, provided that we define the restriction of P to U_i, P_i by $P_i f := P(f\restriction U_i)\restriction U_i$, to have $P_i: S_c(E_1\restriction U_i)\longrightarrow S(E_2\restriction U_i)$. Note that injectivity is not a preserved property under the map $P\longrightarrow P_i$. The local description is now identical to the one for Diff_k, in particular we have the compatibility condition;

$$P_i^{!}\circ t_{ij}^{1} = t_{ij}^{2}\circ P_j^{!} \tag{1}$$

on $\left[S(U_{ij}^{!})\right]^{m_1}$.

Explicitly, we shall say that a continuous operator $P: S(E_1)\longrightarrow S(E_2)$ is in $\mathrm{PDiff}_k(E_1,E_2)$ iff

a/ $\forall$ $f,g\in S(X)$ with disjoint supports, fPg is regularizing, and

b/ in any chart U_i (trivializing for E_1 and E_2), we have $P_i\in \mathrm{PDiff}_k(\mathbb{C}^{m_1},\mathbb{C}^{m_2})$ so that

$$P_i f(x) = \int e^{i\langle x-y,\xi\rangle} a_i(x,\xi) f(y)\,dy\,d\xi$$

with $a_i\in \mathrm{Hom}(\mathbb{C}^{m_1},\mathbb{C}^{m_2})$, a matrix of symbols of order k. In particular we shall assume, that any $P \in PDiff_k(E_1,E_2)$ has homogenous principal symbol $\sigma_k(P)$ of order k, i.e.

$$\sigma_k(P) \in Smbl_k(E_1,E_2) := \{\sigma \in Hom(p_1^*(E_1), p_2^*(E_2)) \mid \sigma(x, r\xi) = r^k \sigma(x,\xi)\}$$

(where $p_1^*(E_1)$ = the induced bundle over $T^*(X) \setminus 0$).

From the general theory of tensor products, we have that given $P \in Diff_k(E_1,E_2)$, $R \in Diff_n(E,E)$, then there exists a <u>unique</u> $P \boxtimes R \in Diff_{k+n}(E_1 \boxtimes E, E_2 \boxtimes E)$, for the external tensor product bundles $E_i \boxtimes E$ $(i=1,2)$ over X x X. We are here interested in $E_i \otimes E = V^*(E_i \boxtimes E)$, the pull-back under the diagonal map $V(x)=(x,x)$. Under this pull-back we loose uniqueness however, as is easily seen, i.e. there is no canonical choice for $P \otimes R$. As we have seen above in the Asada construction, this is exactly the point where connections come in and provide a kind of measure of the degree of non-uniqueness.

For $PDiff_k$, the situation is analogous; under the operation $\ldots \boxtimes E$, there exists a unique operator $P \boxtimes R$, but it needn't be in $PDiff_{k+n}(E_1 \boxtimes E, E_2 \boxtimes E)$ (since the natural external tensor product symbol will in general only be continuous, not smooth). This difficulty is handled by passing over to the closure, $\overline{PDiff_{k+n}(E_1 \boxtimes E, E_2 \boxtimes E)}$. Under $\ldots \otimes E$, as in the case of $Diff_k$, one looses uniqueness, which motivates the construction of Asada connections for $PDiff_k$.

<u>Remark 1</u>: Under $\ldots \otimes E$ there is a natural pairing of symbols for psdo, $P \in PDiff_k(E_1,E_2)$ and $R \in PDiff_n(E,E)$ gives;

$$(\sigma(P) \otimes \sigma(R))(x,\xi) = \sigma(P)(x,\xi) \otimes \sigma(R)(x,\xi)$$

<u>Remark 2</u>: In some sense, the construction of connections here, is analogous to and generalizes the construction in Theorem 3, p. 87 of (PAL).

<u>Remark 3</u>: Tensoring with a trivial $E = X \times \mathbb{C}$, doesn't lead to any complications (trivial obstruction class) in the case of $Diff_k$ (cf. page 123-124 in (AND 1)). The same holds for $PDiff_k$ and shows that the liftings constructed here reflect the vector bundle geometry.

As in the case of Diff_k, we would like to define connection-like objects and we are thus interested in the special kind of liftings $\tilde{P}$ under the operation ... $\otimes E$ such that

$$\sigma_k(\tilde{P}) = \sigma_k(P) \otimes \mathrm{Id}_E. \tag{2}$$

By (1), we have for $f \in [S(U'_{ij})]^{m_1}$ that $P'_i \circ t^1_{ij} f(x) = t^2_{ij} \circ P'_j (f \circ \mu_{ij}) \circ \mu^{-1}_{ij}(x)$, for $x \in U'_{ij}$ and $\mu_{ij} := \mu_i \circ \mu_j^{-1} : U_j \longrightarrow U_i$, the change of coordinates. Writing,

$$P'_i f(x) = \int e^{i\langle x-y,\xi\rangle} a_i(x,\xi) f(y)\,dy\,d\xi$$

we get for $A_{ij} := P'_i \circ t^1_{ij}$ (the composition of two psdo) the asymptotic expansion:

$$\sigma(A_{ij})(x,\xi) \sim \sum_\alpha \frac{1}{\alpha!} a_i^{(\alpha)}(x,\xi) D_x^\alpha t^1_{ij}(x),$$

($a_i^{(\alpha)}(x,\xi) := \partial_\xi^\alpha a_i(x,\xi)$). Similarly, for the right-hand side of (1), which we shall denote by B_{ij}, we have; $B_{ij} f(x) := t^2_{ij}(x) P'_j (f \circ \mu_{ij}) \circ \mu^{-1}_{ij}(x) =$

$$= \int e^{i\langle \mu^{-1}_{ij}(x) - y,\xi\rangle} t^2_{ij}(x) \; a_j(\mu^{-1}_{ij}(x),\xi) f(\mu_{ij}(y))\,dy\,d\xi =$$

$$= \int e^{i\langle \mu^{-1}_{ij}(x) - \mu^{-1}_{ij}(y),\xi\rangle} t^2_{ij}(x) \; a_j(\mu^{-1}_{ij}(x),\xi) \left| \frac{D\mu^{-1}_{ij}}{Dy} \right| f(y)\,dy\,d\xi \tag{3}$$

By a standard argument (cf. (HÖR), p.107-109), one shows that $\sigma(B_{ij})$ has the asymptotic expansion;

$$\sigma(B_{ij})(x,\xi) \sim \sum_\alpha \frac{1}{\alpha!} t^2_{ij}(x) a_j^{(\alpha)}(\mu^{-1}_{ij}(x), {}^t(d\mu_{ij})\xi) \; Y_\alpha(x,\xi).$$

So, the very fact that $P \in \mathrm{PDiff}_k(E_1,E_2)$ is expressed by

$$a_i^{(\alpha)}(x,\xi) D_x^\alpha t^1_{ij}(x) = t^2_{ij}(x) a_j^{(\alpha)}(\mu^{-1}_{ij}(x), {}^t(d\mu_{ij})\xi) \; Y_\alpha(x,\xi) \tag{4}$$

(in the sense of asymptotic equivalence of symbols). $Y_\alpha(x,\xi)$ is here a function, smooth in x and a polynomial of degree $\leq |\alpha|/2$ in ξ.

The liftings $\tilde{P}_i : S(E_1 \otimes E \restriction U_i) \longrightarrow S(E_2 \otimes E \restriction U_i)$ are now defined locally by the property (2);

$\tilde{P}'_i f(x) := \int e^{i\langle x-y,\xi\rangle} (a_i(x,\xi) t^1_{ij} \otimes e_{ij}) f(y)\,dy\,d\xi$, mapping $[S(U'_{ij})]^{m_1+m} \longrightarrow [S(U'_{ij})]^{m_2+m}$.

The Asada connection and the obstruction class is now defined in analogy with the Diff_k-case. Using the same notation as for pdo, we define the asymptotic obstruction $\gamma_{k-1}(W_{ij}) :=$ asymptotic expansion for $(\sigma(\tilde{A}_{ij}) - \sigma(\tilde{B}_{ij}))$, where $\tilde{A}_{ij}f := \tilde{P}_i^! \circ (t^1_{ij} \otimes e_{ij})f$ and $\tilde{B}_{ij}f :=$ $= (t^2_{ij} \otimes e_{ij})\tilde{P}_j^!(f \circ \mu_{ij}) \circ \mu_{ij}^{-1}$. Using (4) we obtain;

$$\gamma_{k-1}(W_{ij})(x,\xi) = \sum_{\alpha} \frac{1}{\alpha!}\left(a_i^{(\alpha)}(x,\xi) \otimes \mathrm{Id}_E\right)\left\{D_x^{\alpha}(t^1_{ij} \otimes e_{ij}) - D_x^{\alpha} t^1_{ij} \otimes e_{ij}\right\}$$

Proceeding inductively, as in the case of Diff_k, we can also define $\gamma_{k-s}(W_{ij})$, the asymptotic obstruction of order $(k-(s+1))$. Clearly, the observations 2°, 3° and 4° for pdo above remain, whereas instead of 1° we have;

$\gamma_{k-s}(W_{ij})$ is a smooth section in the bundle $\mathcal{G}_{k-s} \upharpoonright U_{ij}$, where

$$\mathcal{G}_{k-s} := \mathrm{Smbl}_{k-s}(E_1, E_2) \otimes \mathrm{Hom}(E,E).$$

Similarly, we shall also define the asymptotic obstruction class $\Gamma_{k-s} := \{[\gamma_{k-s}(W_{ij})]\}$ in $H^1(X, \mathrm{Ran}(N^s))$, where N^s is the analogue of M^s,

$$N^s : S(\mathrm{Hom}(E,E)) \longrightarrow \mathrm{Ran}(N^s) \subset S(X, \mathcal{G}_{k-s}),$$

as well as the analogue of the first Chern class $\#^s(\Gamma_{k-s})$ in a 2-cohomology set, i.e.

$$\#^s : H^1(X, \mathrm{Ran}(N^s)) \longrightarrow H^2(X, D^s).$$

Here $D^s :=$ the automorphisms of $S(X, \mathrm{Hom}(E,E))$ which leave $\mathrm{Ker}(N^s)$ invariant.

The associated theory for characteristic classes obtained for psdo in this way, is capable of a much more detailed description of the vector bundle geometry than the polynomial type theory which goes along with pdo. This also has importance for some applications to mathematical physics, which is the subject of a forthcoming paper ((AND 2)). Roughly speaking, this more flexible set of symbols will allow us to introduce proper homotopy relations. For some promising applications, consult (AS 2).

REFERENCES

(AND 1) Andersson, S.I. ; Vector Bundle Connections and Liftings of Partial Differential Operators. Lecture Notes in Math., Vol 905, p.119-132 (1982).

(AND 2) Andersson, S.I. ; Gauge Theory as a Low-Dimensional Non-abelian Hodge Theory, seminar in the Jânos Bolyai Math. Soc., Budapest (1983), to appear.

(AS 1) Asada, Akira ; Connections of Differential Operators, J. Fac. Sci., Shinshu Univ., Vol. 13, 87-102 (1978).

(AS 2) Asada, Akira ; Curvature Forms with Singularities and Non-Integral Characteristic Classes, Invited talk at the XII Int. Conf. on Differential Geometric Methods in Theor. Physics, Clausthal Aug.-Sept. 1983 (to appear in the proceedings).

(DE) Dedecker, P. ; Sur la cohomologie non-abélienne I, Canad. J. Math. 12 ,231-251, (1960).

(HÖR) Hörmander, L. ; Fourier Integral Operators I, Acta Math. 79-183, 127 (1971).

(PAL) Palais, R.S. ; Seminar on the Atiyah-Singer Index Theorem, Ann. Math. Studies 57, Princeton 1965.

EUCLIDEAN YANG-MILLS FLOWS IN THE ORBIT SPACE

M. Asorey
Departamento de Física Teórica.
Facultad de Ciencias.
Universidad de Zaragoza.
Spain.

ABSTRACT

We establish a global setting for the canonical formalism of the Euclidean Yang-Mills theory in the orbit space M. In this setting it is shown that the Euclidean Yang-Mills equations lead to an ordinary second order differential system of M and the (anti) self-dual solutions are in the flow of a densely defined vector field $\pm H$ of M. We also prove that, for the particular case of having T^3 as space manifold the flows $\pm H$ are homotopic complete, i.e., in each class of $\Pi_1(M)$ there exists a closed integral curve of $\pm H$.

1.- INTRODUCTION.

In the last two years, some progress has been done in the construction of the regularized Quantum Yang-Mills theory from the continuum point of view in the orbit space [3,11]. On the other hand, classical solutions of Euclidean Yang-Mills Equations with finite Euclidean action are expected to play a relevant role in the quantum theory.

The aim of the present paper is to set up an intrinsic and global description of these solutions in the orbit space in order to study their possible relevance for the regularized quantum theory.

2.- EUCLIDEAN YANG-MILLS EQUATIONS.

Let (V,h) be a connected, compact, oriented, 3-dimensional Riemannian manifold and G a connected, simply connected, simple, compact, Lie group. Let $P(V\times S^1,G)$ by any principal fibre bundle with base manifold $V\times S^1$ and structure group G. The equivalence classes of such principal bundles are in one-to-one correspondence with the elements of $H^4(V\times S^1, \mathbb{Z})$ [6]. Since $H^4(V\times S^1, \mathbb{Z}) = \mathbb{Z}$, those equivalence classes are parametrized by an integer number. In the case G = SU(N) this number is given by $-c_2(P)$, where $c_2(P)$ denotes the second Chern class of P.

Definition 2.1.- The Yang-Mills fields on $V \times S^1$ are the smooth connections B defined in such principal bundles $P(V\times S^1, G)$.

We denote by B both the connection and the 1-form of the connection. Let * be the Hodge star operator with respect to the Euclidean metric of $V\times S^1$ given by the direct sum of h and the metric $d\theta^2$ of S^1.

Definition 2.2.- The Euclidean Yang-Mills dynamics is provided by the stationary connections of the functional

$$S_E(B) = -\frac{1}{2}\int_{V\times S^1} \mathrm{Tr}\ F(B)\wedge{*F(B)} \qquad (2.1)$$

where $F(B) = dB + 1/2\,[B,B]$ is the curvature 2-form of B, and $\mathrm{Tr}\ (\sigma_1\sigma_2) = \mathrm{tr}\ (\mathrm{Ad}\ \sigma_1\circ\mathrm{Ad}\ \sigma_2)$ for any $\sigma_1, \sigma_2 \in$ Lie G.

We recall that differential characterizations of the stationary points of (2.1) are provided by the following two propositions.

Proposition 2.3.- The stationary points of (2.1) are solutions of the differential equations

$$d_B\ {*F(B)} = 0 \quad \text{Euclidean Yang-Mills equations} \qquad (2.2)$$

where d_B denotes the covariant differential with respect to B.

Besides, F(B) satisfies the equation

$$d_B\ F(B) = 0 \ \text{(Bianchi identity)} \qquad (2.3)$$

Proposition 2.4.- The Yang-Mills fields B satisfying the equation

$$*F(B) = \pm\ F(B) \qquad \text{(anti) self-duality} \qquad (2.4)$$

are particular solutions of (2.2).

The proof follows trivially from (2.3) and (2.4).

Definition 2.5.- The smooth $V\times S^1$ automorphisms of P are called gauge transformations.

Proposition 2.6.- The functional (2.1) and the differential equations (2,2)-(2.4) are invariant under gauge transformations.

3.- GLOBAL SETTING FOR THE CANONICAL FORMALISM.

This section is devoted to set up an intrinsic framework for the canonical formalism of Yang-Mills theory wich does not depend on whether P is trivial or not. In fact, we develop the framework used in [2], [4] for different purposes.

Let B be an arbitrary connection of P. We denote by $P_\theta(V\times\{\theta\}, G)$ to the principal fibre bundle obtained by restriction of P to $\Pi^{-1}(V\times\{\theta\})$. For any $\theta\in S^1$, $P_\theta(V,G)$ is trivial because G is simply connected and $\dim V = 3$.

Let

$$\tau_\theta^B : P_0 \longrightarrow P_\theta \tag{3.1}$$

be the map defined by means of the parallel transport with respect to B along the curves

$$c_x : [0,\theta] \longrightarrow V\times S^1$$

such that $c_x(\psi) = (x,\psi)$, $0 \leqslant \psi \leqslant \theta$.

Proposition 3.1.- The map (3.1) is an isomorphism of principal fibre bundles.

Proof: It is an immediate consequence of the properties of parallel transport [7].

Let $B(\theta)$ be the connections of P_θ defined by restriction of B to P_θ.

Definition 3.2.- We define the family $A(\theta)$, $0 \leqslant \theta \leqslant 2\pi$ of connections of $P_0(V,G)$ by pull-back from $B(\theta)$ by means of τ_θ^B ; i.e., $A(\theta) = \tau_\theta^{B*}B(\theta)$.

Proposition 3.3.- The family of connections $A(\theta)$ satisfies the following boundary conditions.

$$A(0) = B(0) \tag{3.2a}$$

$$A(2\pi) = A(0)^{\tau_{2\pi}^B} \tag{3.2b}$$

where $\tau_{2\pi}^B$ is the automorphism of P_0 given by (3.1).

Proof: (3.2a) follows from definition (3.2) and (3.2b) from the boundary property $B(2\pi) = B(0)$ of $B(\theta)$.

If $\mathcal{B}$ denotes the affine space of smooth connections of P and $\mathcal{A}$ the space of those of P_0, let

$$K : \mathcal{B} \longrightarrow \text{Map}\ ([0,1],\mathcal{A}) \qquad (3.3)$$

be the map given by definition (3.2)

4.- CANONICAL YANG-MILLS EQUATIONS.

Hereafter we will embed the space of smooth (C^∞) connections $\mathcal{A}$ in the space of Sobolev connections $\mathcal{A}_K$ with $K \geq 5$ (see [5,8] for the definitions Since $K \geq 5$, the Sobolev's lemma implies that any connection $A \in \mathcal{A}_K$ is a C^2-connection of P_0. $\mathcal{A}$ is a dense subset of $\mathcal{A}_K$. Once a connection $A°$ in $\mathcal{A}_K$ is fixed, $\mathcal{A}_K$ can be identified with the Hilbert space of Sobolev sections $\Gamma_K(\text{ad}P_0 \otimes \Lambda^1 V)$ of the vector bundle $\text{ad}P_0 \otimes \Lambda^1 V$. Let $\|\ \|_K$ denote the Hilbert norm of $\Gamma_K(\text{ad}P_0 \otimes \Lambda^1 V)$, [8]. Since P_0 is trivial we choose a trivial connection as background connection $A°$.

Proposition 4.1.- A connection $B \in \mathcal{B}$ is a solution of the Euclidean Yang-Mills equations (2.2) if and only if the corresponding image through K, $K(B) = (A\ (\theta))$ satisfy the equations

$$* d_A * \dot{A} = 0 \quad \text{Gauss law} \qquad (4.1a)$$

$$* d_A * F(A) = \ddot{A} \quad \text{Newton law} \qquad (4.1b)$$

where $\dot{A}$ denotes the time derivative $\frac{d}{d\theta} A(\theta)$ with respect the norm $\|\ \|_K$ of $\mathcal{A}_K$.

Proof: Let σ be a cross section of $P_0(V,G)$ and $\Sigma : V \times S^1 \longrightarrow P$ the map defined by $\Sigma(x,\theta) = \tau^B_\theta \sigma(x)$. Σ defines a local cross section of P; i.e. for any $(\theta_1,\theta_2) \subset S^1$, Σ is a cross section of the principal fibre bundle $\Pi^{-1}(V \times (\theta_1 \theta_2))$.

In this particular local section of P,

$$B^\Sigma(x,\theta) = A^\sigma(\theta)(x) \oplus 0 \qquad (4.2)$$

when $B^\Sigma : \Sigma^*(B)$, $A^\sigma(\theta) = \sigma^*(A(\theta))$ and 0 is the trivial 1-form of (θ_1,θ_2)

Then it is trivial to check that the expression of (2.2) in this local section is the same as that of (4.1a,b). Because of the local character of eqs. (2.2) and (4.1 a,b), this fact is sufficient to prove the proposition.

In a similar way one proves:

Proposition 4.2.- A necessary an sufficient condition for a Yang-Mills field B satisfies the (anti) self-duality equations (2.4) is that the corresponding image through K, $(A(\theta))$ verifies the equation

$$\dot{A} = (-) *F(A) \tag{4.3}$$

In the same way the solutions of the Yang-Mills equations corresponding to the canonical Lorentz metric of $V\times S^1$ verify the equations (4.1a) and

$$\ddot{A} = - *d_A*F(A) \tag{4.4}$$

Equations (4.1b) and (4.4) are second order ordinary differential equations as Newton equation of motion. In both cases the term $*d_A*F(A)$ plays the role of a Yang-Mills force and the only difference between both cases is the sign of this force.

5.- YANG-MILLS EQUATION AND (ANTI) SELF-DUAL FLOWS IN THE ORBIT SPACE.

The right hand side of equations (4.1) and (4.3) define two tensorial one-formes of P_0 which transform under any gauge transformation τ of P as follows

$$*d_{A^\tau}*F(A^\tau) = \tau\,(*d_A*F(A))\,\tau^{-1}$$

$$*F(A^\tau) = \tau\,(*F(A))\,\tau^{-1}$$

Therefore $*d_A*F(A) \in \Gamma_{k-2}(\mathrm{ad}\,P \otimes \Lambda^1 V)$ and $*F(A) \in \Gamma_{k-1}(\mathrm{ad}\,P_0 \otimes \Lambda^1 V)$.

If $A-A_0 \in \Gamma_{k+1}$, $*F(A) \in \Gamma_k$ and for $A-A_0 \in \Gamma_{k+2}$, $*d_A*F(A) \in \Gamma_k$. Hence, in the respective cases $*d_A*F(A)$ and $*F(A)$ define tangent vectors at A of A_k. Let $\tilde{\mathcal{F}}$ and $\tilde{\mathcal{H}}$ be the corresponding densely defined vector fields. $\tilde{\mathcal{F}}$ is only defined at the points with $A-A_0 \in \Gamma_{k+2}$ and $\tilde{\mathcal{H}}$ at those with $A-A_0 \in \Gamma_{k+1}$. Then the solutions of the Euclidean Yang-Mills equations are the integral curves (spray) of the second order differential equation governed by $\tilde{\mathcal{F}}$ which satisfy the constraint equation (4.1a), and the (anti) self-dual solutions are the integral curves (flow) of

$(-)\tilde{H}$. However, the existence of solution to the Cauchy problem of both differential equations can not be directly implemented because of the densely defined character of $\tilde{F}$ and $\tilde{H}$.

There are two reasons for improve the above canonical setting of Euclidean Yang-Mills equations. First, A_k does not provide the minimal setting because there is a lot of gauge transformations which leave the equations (4.1a,b) and (4.3) invariant. In fact, they are invariant under any gauge transformation of P_0. Therefore it suffices to know only one solution in each equivalence class. Furthermore, since the appropriate framework for the quantum theory is not A_k but the space of gauge equivalence classes of A_k (orbit space) [3,4,11], it is suitable to set the solutions of the Euclidean Yang-Mills equations in the orbit space framework in the order to study their quantum effects.

Let G^0_{k+1} be the group of gauge transformations of P_0 in Sobolev class k+1 which leave fixed a given point u_0 of P_0. Then we have

<u>Proposition 5.1</u>. [4,9,10] .- The orbit space $M = A_k/G^0_{k+1}$ is a smooth Hilbert manifold and $A_k(M,G^0_{k+1})$ is a principal fibre bundle with structure Lie group G^0_{k+1}.

Let $\tilde{g}$ be the smooth, gauge invariant, Riemannian metric of A_k defined in [3]. Let V_a be the closed subspace of $T_A A_k$ tangent to the fiber through A, and H_A that defined by

$$H_A = \mathrm{Ker} *d_A* \oplus (V_A \oplus \mathrm{Ker} *d_A*)^{\perp} \qquad (5.1)$$

where $\perp$ means orthogonality with respect to $\tilde{g}$.

<u>Proposition 5.2</u>. (modified Coulomb connection).- The distribution H_A defines a connection in the principal fibre bundle $A_k(M,G^0_{k+1})$.

Proof: It is clear that H_A is closed and

$$T_A A_k = H_A \oplus V_A \qquad (5.2)$$

is a splitting of $T_A A_k$. H_A is a smooth distribution. Moreover, $\tilde{g}$ is G^0_{k+1}-invariant and

$$\mathrm{Ker} *d_{A\tau}* = R(\tau)_* (\mathrm{Ker} *d_A*)$$

where $R(\tau)$ means the right action of $\tau \in G^0_{k+1}$ on A_k. Hence

$$H_{A\tau} = R(\tau)_* H_A$$

which proves the proposition.

The subspace $(V_A \oplus \text{Ker} *d_A*)^{\perp}$ is finite dimensional and its dimension is less than dim G .

Proposition 5.3.- $\tilde{H}$ is a horizontal vector field with respect to ω_0. Furthermore, any solution of the canonical Yang-Mills Equations (4.1 a,b) is horizontal with respect to ω_0.

Proof: Let $A(\theta)$ be any integral curve of $\tilde{H}$. Then

$$\dot{A} = *F(A)$$

and $$*d_A*\dot{A} = *d_A**F(A) = *d_A\ F(A) = 0$$

which vanishes because of Bianchi's identity $d_A\ F(A) = 0$. Hence $\dot{A}(\theta) \in \text{Ker} *d_A* \subset H_A$. In the same way any solution of the canonical Euclidean Yang-Mills equations satisfies Gauss law (4.1a) which implies that $\dot{A}(\theta) \in \text{Ker} *d_A* \subset H_A$.

Since $\tilde{F}$ and $\tilde{H}$ are gauge invariant, there exist two densely defined vector fields F and H of M such that

$$F = \Pi_*\tilde{F} \qquad , \qquad H = \Pi_*\tilde{H} \tag{5.3}$$

Accordingly, the orbits in M of the solutions of equations (4.1 a,b) lie in the integral curves of the spray of the second order differential equation governed by F; i.e.,

$$d^2[A(\theta)]/d\theta^2 = F(A(\theta)) \tag{5.4}$$

where $[A(\theta)]$ denotes $\Pi(A(\theta))$. The orbits of the (anti) self-dual solutions (4.3) are in the flow of the vector field $(-)H$; i.e.,

$$d[A(\theta)]/d\theta = (-)H(A(\theta)). \tag{5.5}$$

Conversely, we have:

Corollary 5.4.- The solutions of the equations (4.1 a,b) and (4.3) are horizontal lifts with respect to ω_0 of curves satisfying (5.4) and (5.5), respectively.

The proof is a straightforward consequence of Proposition (5.3).

The critical points of the vector fields F and H are given by the orbits of connections A with $d_A{*F(A)} = 0$ and ${*F(A)} = 0$ respectively.

The points with $H_{[A]} = 0$ (${*F(A)} = 0$) give rise to static solutions of Yang-Mills equations. They are the only points where the (anti) self-dual solutions can be degenerate. When $V = S^3$ there is only one critical point of H in M. However, for general manifolds V the set of such points is not trivial. In [1] we compute explicitely that set for many different manifolds V and structure groups G.

Now, since the closed integral curves of the equation (5.4) are destinated to play a relevant role for the quantum theory we focus our attention on the study of such solutions.

Proposition 5.5.- For any solution B of the Euclidean Yang- Mills equations (2.2) with $\tau^B_{2\pi}(u_0) = u_0$, the induced solution of (5.4) is a closed curve in M.

Proof: Since $\tau^B_{2\pi} \in G^0_{k+1}$ the proof is trivial, because the curve $K(B) = (A(\theta))$ satisfies (4.1 a,b) and the boundary condition (3.2 b). Hence $[A(0)] = [A(2\pi)]$.

In fact, it can be shown that all the closed solutions of (5.4) can be obtained in this way from the solutions of (2.2) in $P(V\times S^1,G)$ by considering in S^1 different Riemannian structures.

The following proposition shows that the set of such solutions is not empty and topologically non-trivial.

Proposition 5.6. (Homotopic completness).- If $G = SU(N)$ with $N>1$, and $V = T^3$ the (anti) self-dual flow has closed solutions in all the homotopy classes of $\Pi_1(M)$.

Proof: In [4] we have proved that the closed curves of A_k induced through K from the connections of a principal fibre bundle $P(V\times S^1,G)$ are homotopic equivalent. Moreover, the correspondence induced in such a way between the classes of fibre bundle $P(V\times S^1,G)$ and the elements of $\Pi_1(M)$ is bijective [4]. Therefore it suffices to find a self-dual or (anti)self-dual connection B in each class of principal fibre bundles $P(T^4, SU(N))$ with $\tau^B_{2\pi}(u_0) = u_0$ for a given point $u_0 \in P_0 \subset P$.

Let σ be a non trivial element of Lie SU(N) such that $\mathrm{tr}\,\sigma^2 = -1$. Then we define the connections

$$B_n = (n/8\pi^2)(\theta_0 \wedge d\theta_1 + \theta_2 \wedge d\theta_3)\,\sigma \qquad n \in \mathbb{N}$$

which obviously verify that $\tau^{B}_{2\pi}(u_0) = u_0$.

It is trivial to see that B_n is self-dual and that n is its corresponding Pontrjagin number:

$$-(1/8\pi^2)\int \mathrm{tr}\, F \wedge F = n$$

Therefore B_n is defined in a principal bundle $P(T^4, SU(N))$ of the correponding Pontrjagin class. In the same way

$$\bar{B}_n = (n/8\pi^2)(\theta_0 \wedge d\theta_1 - \theta_2 \wedge d\theta_3)\,\sigma \qquad n \in \mathbb{N}$$

defines an (anti) self-dual connection in the principal bundle $P(T^4, SU(N))$ with negative Pontrjagin number -n, and such that $\tau^{\bar{B}_n}_{2\pi}(u_0) = u_0$. The proof follows from the fact that all integers can be obtained in such a way.

In fact, the above proposition can be generalized for more general structure groups G.

Proposition 5.7.- If $V = T^3$ for any connected, simply connected, simple, compact Lie group G the (anti) self-dual flow is homotopic complete in M.

Proof: It is a straightforward consequence of Proposition 5.6 and the fact that any principal fibre bundle $P(V\times S^1, G)$ with topological index n in $H^4(V\times S^1, \mathbb{Z})$ is reducible to a SU(2) principal fibre bundle with the same topological index, [5]. Then, the connections B_n and $\bar{B}_n$ defined in the proof of proposition 5.6 can also be considered as connections of the corresponding principal fibre bundles $P(V\times S^1, G)$ and the associated closed curves in M lie in all homotopy class of $\Pi_1(M)$.

ACKNOWLEDGEMENT.-

I wish to thank to P. K. Mitter for stimulating conversations which have enlightened many aspects of this paper.

REFERENCES

[1] M. Asorey. J. Math. Phys. 22, 179 (1981).

[2] M. Asorey, J.F. Cariñena, M. Paramio. J. Math.Phys. 23, 1451 (1982)

[3] M. Asorey, P.K. Mitter. Commun. Math. Phys. 80, 43 (1981)

[4] M. Asorey, P.K. Mitter. "On geometry, topology and θ-sectors in regularized quantum Yang-Mills theory". CERN preprint (1982)

[5] M.F. Atiyah, N.Hitchin, I.M. Singer. Proc. Roy. Soc. A 362, 425 (197

[6] C.J. Isham, in "Recent Developments in Gauge Theories" Plenum, N.Y. (1979).

[7] S. Kobayashi, K. Nomizu "Foundations of Differential Geometry" Vol.1, Wiley, N.Y. (1963).

[8] P.K. Mitter, C.M. Viallet, Commun. Math. Phys. 79, 457 (1981).

[9] M.S. Narasimhan, J.R. Ramadas. Commun.Math. Phys. 67, 121 (1979).

[10] I.M. Singer. Commun. Math. Phys. 60, 7 (1978).

[11] I.M. Singer. Physica Scripta 24, 817 (1981).

CONGRUENCE, CONTACT ET REPÈRES DE FRENET

D. Bernard
I.R.M.A. Laboratoire associé au C.N.R.S.
7, Rue René Descartes
67084. Strasbourg. FRANCE.

1.- Introduction

Soit M une variété où opère différentiablement un groupe de Lie G. Deux sous-variétés V et $\tilde{V}$ de M sont G-congruentes s'il existe $A \in G$ tel que $A(V) = \tilde{V}$.

Le problème de G-congruence des sous-variétés est celui-ci: trouver des conditions infinitesimales sur V et $\tilde{V}$ permettant d'affirmer qu'elles sont G-congruentes.

La méthode classique consiste à construire, pour chaque entier $p < \dim M$, des invariants différentiels des sous-variétés de dimension p de M sous l'action du groupe G et d'essayer d'en construire assez pour que l'egalité de valeurs prises par ces invariants sur deux sous-variétés V et $\tilde{V}$ assure leur G-congruence.

Dans cet article, nous analysons cette méthode d'aprés les travaux de G.R. JENSEN, [7] et S. HUCKEL [8] et nous en montrons les limites: les sous-variétés auxquelles elle s'applique doivent satisfaire des conditions très fortes de régularité.

Pour s'affranchir de ces limites, S. HUCKEL a étudié le problème de G-congruence en termes de G-contact d'ordre k (§4 ci-dessous). Elle a mis en évidence le rôle joué par les propriétés géométriques de la "variété des G-contacts d'ordre k de V avec $\tilde{V}$" (§6) et obtenu un nouveau théorème de congruence dont le théorème classique (E. CARTAN, [4], JENSEN, [7]) est un corollaire. Ce théorème ne suppose pas que l'action de G sur M soit transitive et il s'applique à des sous-variétés *a priori* exclues du champ d'application du théorème classique. C'est le cas des surfaces possédant un ombilic isolé dans un espace euclidien de dimension trois, G étant le groupe des déplacements: nous montrons ici comment le théorème de S. HUCKEL permet de les classifier à G-congruence près dans les cas génériques.

Toutes les donnés: variété M, actions du groupe, sous-variétés V, $\tilde{V}$,... sont supposées C^∞. L'opération du groupe G sur M est notée

$$(g,x) \longmapsto gx \qquad g \in G \qquad x \in M$$

L'espace tangent en x à une variété V est noté V_x on T_xV. Nous notons enfin $n = \dim M$.

2.-Exemples

Soit $M = E^3$ un espace euclidien de dimension 3 et $G = E(3)$ le groupe des déplacements de M.

a) Soit C une courbe orientée de M. On sait en construire des invariants différentiels à condition qu'elle soit birégulière: ce sont la courbure c, la torsion τ et leurs dérivées successives par rapport à une abscisse curviligne s.

Le théorème élémentaire de congruence de deux courbes orientées birégulières C et $\tilde{C}$ peut s'énnoncer ainsi:

Théorème A.- Pour que C et $\tilde{C}$ soient G-congruentes, il faut et suffit qu'il existe un difféomorphisme $\phi: C \longrightarrow \tilde{C}$ tel que

$$\phi^* d\tilde{s} = ds \qquad \phi^* \tilde{c} = c \qquad \phi^* \tilde{\tau} = \tau$$

Sous des hypothèses restrictives, on en déduit (cf. [8]) des théorèmes de congruence qui s'expriment uniquement en termes d'invariants différentiels. Le plus simple est celui-ci

Théorème B.- Soit C, une courbe birégulière sur laquelle $c' = \frac{dc}{ds}$ ne s'annule pas. Pour qu'une courbe birégulière $\tilde{C}$ soit G-congruente à C il faut et suffit que les deux applications

$$(c,c',\tau) : C \longrightarrow \mathbb{R}^3 \quad \text{et} \quad (\tilde{c},\tilde{c}',\tilde{\tau}) : \tilde{C} \longrightarrow \mathbb{R}^3$$

aient la même image; c'est-à-dire que C et $\tilde{C}$ aient "globalement les mêmes invariants d'ordre 3".

A la différence du théorème A, ce théorème ne prescrit pas à l'avance une correspondance entre C et $\tilde{C}$.

b) Pour les surfaces de E^3 , il y a un théorème de congruence élémentaire, le théorème de BONNET:

Théorème C.- Les surfaces connexes orientées V et $\tilde{V}$ de E^3 sont G-congruentes si et seulement si il existe un difféomorphisme $F : V \longrightarrow \tilde{V}$ tel que

$$F^* \tilde{I} = I \qquad F^* \tilde{II} = II$$

ou I et II (resp. $\tilde{I}$ et $\tilde{II}$) sont les formes quadratiques fondamentales de V et $\tilde{V}$.

Les invariants différentiels d'ordre 2 d'une surface orientée V sont sa courbure totale K et sa courbure moyenne H, ou ce qui est equivalent, la couple de ces courbures principales k_1 , k_2 $(k_1 \geq k_2)$. Cependant les fonctions H et K sont C^∞ , tandis que les fonctions k_i ne sont pas différentiables en des ombilics isolés de V .

On obtient les invariants d'ordre 3 en prenant les dérivées de Lie $k_{i,j} = L_{e_i} k_j$ des fonctions k_j par rapport a des champs locaux de vecteurs unitaires de directions principales $\{\vec{e}_i\}$ ($\vec{e}_i$ associé à k_i; $(\vec{e}_i, \vec{e}_j)$ compatible avec l'orientation); l'existence de tels champs locaux C^∞ suppose encore que V n'ait pas d'ombilic ou, au contraire, que tous ces points soient des ombilics.

Du théorème de Bonnet, on déduit facilement des théorèmes de congruence analogues au théorème B, par exemple celui-ci

Théorème D.- Soient V et $\tilde{V}$ des surfaces simplement connexes orientées. On suppose de plus que V n'a pas d'ombilic et que les fonctions k_1 , k_2 sont partout indépendantes $(dk_1 \wedge dk_2\ (p) \neq 0$ pour tout $p \in V)$. Pour que $\tilde{V}$ soit (localement) G-congruente à V il faut et suffit qu'il existe $\varepsilon = \pm 1$ tel que

$$(k_1, k_2, k_{1,1}, \ldots, k_{2,2})(V) = (\tilde{k}_1, \tilde{k}_2, \varepsilon\tilde{k}_{1,1}, \ldots, \varepsilon\tilde{k}_{2,2})(\tilde{V}) ,$$

la G-congruence étant assurée au voisinage de toute couple de points $x \in V$, $\tilde{x} \in \tilde{V}$ où les invariants des deux surfaces son égaux.

Les autres théorèmes de ce type supposent tous que $dk_1 \wedge dk_2 = 0$ sur tout V, le cas où il existe des points <u>isolés</u> où $(dk_1 \wedge dk_2)(p) = 0$ étant totalement exclu du champ de la théorie.

A l'opposé, le cas où tous les points de V sont des ombilics est très simple:

<u>Théorème E.- Soit V un ouvert d'une sphère ou d'un plan. Pour qu'une surface $\tilde{V}$ soit localement congruente à V, il faut et suffit que</u>

$$(k_1,k_2)(V) = \pm (\tilde{k}_1,\tilde{k}_2)(\tilde{V}) .$$

3.- Eléments de contact

a) <u>contact d'ordre h de sous-variétés</u>.- Soient V et $\tilde{V}$ deux sous-variétés de M de dimension $p < n$. Elles ont en x_o un contact d'ordre zero si $x_o \in V \cap \tilde{V}$. Pour définir le contact d'ordre $h \geqslant 1$, supposons que $V = f(S)$ où $f : S \longrightarrow M$ est un plongement -nous noterons alors $V = (S,f)$- et que $\tilde{V} = (\tilde{S},\tilde{f})$. Les sous-variétés V et $\tilde{V}$ ont en $x_o = f(s_o)$ et $\tilde{x}_o = \tilde{f}(s_o)$ un contact d'ordre $k > 0$ si et seulement s'il existe un difféomorphisme local $\phi : (S,s_o) \longrightarrow (\tilde{S},\tilde{s}_o)$ tel que

$$j^k_{s_o} f = j^k_{\tilde{s}_o} \tilde{f} \cdot j^k_{s_o} \phi$$

La définition ne dépend pas du choix particulier des paramétrages (S,f) et $(\tilde{S},\tilde{f})$; on voit facilement que pour $h \geqslant 1$, elle équivaut à la propriété suivante:

Soit $O \subset M$ un voisinage ouvert de x_o muni de coordonnées locales $(x,y) = (x^1,\ldots,x^p,y^1,\ldots y^{n-p})$ nulles en x_o, dans laquelle V est un graphe

$$O \cap V = \{(x,y) \in O \;/\; x \in U \;,\; y = F(x)\}$$

où U est un voisinage ouvert de 0 dans $\mathbb{R}^p$ et F une application C^∞ de U dans $\mathbb{R}^{n-p}$.

Alors, une sous-variété $\tilde{V}$ a en x_o un contact d'ordre $h \geqslant 1$ avec V si et seulement si:

i) dans un ouvert O', $x_0 \in O' \subset O$, la sous-variété $\tilde{V}$ est elle-même un graphe pour le même système de coordonnées locales :

$$O' \cap \tilde{V} = \{(x,y) \in O' \;/\; x \in U' \;,\; y = \tilde{F}(x)\}$$

où U' est un voisinage de O dans $\mathbb{R}^p$ et $\tilde{F}$ une application C^∞ de U' dans $\mathbb{R}^{n-p}$.

ii) F et $\tilde{F}$ ont mêmes dérivées d'ordre $\leqslant h$ en 0:

$$D^\ell F(0) = D^\ell \tilde{F}(0) \qquad \ell = 0,1,2,\ldots,h.$$

b) on peut définir les éléments de contact de dimension p comme classes d'équivalence de jets d'immersions de $(\mathbb{R}^p,0)$ dans M (cf. AMBROSE [1]). Il sera plus commode ici de les définir, selon JENSEN, comme des élements "holonomes" d'un espace plus grand que celui des éléments de contact.

Fixons un entier $p < n$ et notons $G_p(M)$ la variété des p-plans tangents à M. L'application canonique $\alpha: G_p(M) \longrightarrow M$ est une fibration dont la fibre type est la grassmannienne des sous-espaces vectoriels de dimension p de $\mathbb{R}^n$. On définit par itération les <u>fibrés en grassmanniennes successifs</u> sur M en posant

$$G_p^0(M) = M \;;\qquad G_p^{k+1}(M) = G_p(G_p^k(M)),\qquad k \in \mathbb{N}.$$

Nous noterons $\alpha_k: G_p^k(M) \longrightarrow G_p^{k-1}(M)$ la fibration canonique.

Tout difféomorphisme ϕ de M sur M s'étend à $G_p(M)$ en posant, pour $y \in G_p(M)$,

$$\phi(y) = \phi^T_{\alpha(y)}(y)$$

ce qui implique $\alpha(\phi(y)) = \phi(\alpha(y))$. De proche en proche, ϕ s'étend aux grassmanniennes successives et l'on a pour tout entier $k \geqslant 1$

$$\alpha_k \circ \phi = \phi \circ \alpha_k$$

En particulier, si le groupe de Lie G opère sur M, pour tout $g \in G$ l'application $L_g: x \longrightarrow gx$ est un difféomorphisme de M et s'étend à $G_p^k(M)$ pour tout entier k. Ceci permet d'étendre à $G_p^k(M)$ l'action de G en posant

$$g\,y = L_g(y) \qquad \text{pour } g \in G \text{ et } y \in G_p^k(M),\quad k \geqslant 1.$$

c) Prolongement des immersions.

Soit $f: S \longrightarrow M$ une immersion et $p = \dim S$. Pour tout $s \in S$, l'image $f_s^T(S_s)$ est un p-plan tangent à M en $f(s)$. On peut ainsi définir une application

$$T f : S \longrightarrow G_p^1(M)$$

en posant $T f(s) = f_s^T(S_s)$. On a alors $\alpha \circ Tf = f$ ce qui implique que Tf est encore une immersion et que le procedé peut être iteré en posant

$$T^0 f = f, \quad T^h f = T(T^{h-1} f) \qquad k \geq 1.$$

On définit ainsi pour tout entier k une immersion

$$T^k f : S \longrightarrow G_p^k(M)$$

qui est un prolongement d'ordre k de f.

Pour tout entier $k \geq 1$, les propriétés suivantes sont immédiates

(1) $\alpha_k \circ T^k f = T^{k-1} f$

$\phi \circ T^k f = T^k(\phi \circ f)$ si ϕ est un difféomorphisme de M.

(2) $L_g \circ T^k f = T^k(L_g \circ f)$, pour tout $g \in G$

(3) $T^k(f \circ F) = (T^k f) \circ F$, si $F: \tilde{S} \longrightarrow S$ est un difféomorphisme.

On montre alors:

Proposition 1.- Les sous-variétés $V = (S,f)$ et $\tilde{V} = (\tilde{S},\tilde{f})$ ont en $x_0 = f(s_0)$ et $\tilde{x}_0 = \tilde{f}(x_0)$ un contact d'ordre $k \geq 0$ si et seulement si

$$T^k f(s_0) = T^k \tilde{f}(\tilde{s}_0)$$

De plus la propriété (3) montre que si (S,f) et $(\tilde{S},\tilde{f})$ sont deux parametrages d'une même sous-variété V - ce qui implique $\tilde{f} = f \circ F$ où $F: \tilde{S} \longrightarrow S$ est un difféomorphisme - des qui $\tilde{f}(\tilde{s}) = f(s)$, on a aussi $T^k \tilde{f}(\tilde{s}) = T^k f(s)$, propriété qui s'étend immédiatement aux parametrages locaux. Ces remarques permettent d'identifier l'élément de contact d'ordre k de V en $x_0 = f(s_0)$ à l'élément $T^k f(s_0)$ de $G_p^k(M)$; la variété $C_p^k(M)$ des éléments de contact d'ordre k et de dimension p de M s'identifie à une sous-variété de $G_p^k(M)$ et il découle de (2) qu'elle est stable par l'action de G sur $G_p^k(M)$.

Notons enfin que si $y \in G_p^K(M)$ et $k \geqslant 2$ on a

$$\alpha_k(y) = \alpha_{k-1}^T(y)$$

mais cette propriété ne suffit pas à caractériser $G_p^k(M)$, même si $k = 2$, (cf. [1]).

4.- G-Contact.

a) Définition 1. Soit M une variété sur laquelle opère un groupe de Lie G, On dit que deux sous-variétés V et $\tilde{V}$ de M ont en $x \in V$ et $\tilde{x} \in \tilde{V}$ un "G-contact d'ordre k" s'il existe $g \in G$ tel que $gx = \tilde{x}$, les sous-variétés $g(V)$ et $\tilde{V}$ ayant en $\tilde{x}$ un contact d'ordre k.

Si V et $\tilde{V}$ sont parametrées par (S,f) et $(\tilde{S},\tilde{f})$, il résulte de la proposition 1 et de la relation (2) que

Proposition 2.- Les sous-variétés V et $\tilde{V}$ ont en $x = f(s_0)$ et $\tilde{x} = \tilde{f}(\tilde{s}_0)$ un G-contact d'ordre k si et seulement si il existe $g \in G$ tel que

$$gT^k f(s_0) = T^k \tilde{f}(\tilde{s}_o) \tag{4}$$

c'est-à-dire, si leurs éléments de contact d'ordre k en x et x appartiennent à la même orbite de $G_p^k(M)$ sous l'action de G.

b) Si V et $\tilde{V}$ sont G-congruentes avec $\tilde{V} = A(V)$, la condition précédent est vérifiée pour tous couples de points $x \in V$ et $\tilde{x} = Ax$, et ceci pour tout entier k. Une condition nécessaire de G-congruence est donc que V et $\tilde{V}$ aient un "G-contact global" de tous les ordres au sens de la définition suivante:

Définition 2.- On dit que les sous-variétés $V = (S,f)$ et $\tilde{V} = (\tilde{S},\tilde{f})$ ont un contact global d'ordre k si

$$GT^k f(S) = GT^k \tilde{f}(\tilde{S})$$

Il est remarquable que, dans les cas les plus simples, la condition très grossière de G-contact global d'ordre k, pour k petit, soit suffisante pour assurer la G-congruence. Ainsi dans les exemples de §2, les hypothèses sont en fait des hypothèses de G-contact global

Théorème B.- Si C est birégulière et si dc/ds ne s'annule pas sur C, il suffit que C' ait avec C une G-contact global d'ordre 3 pour assurer la G-congruence de C et C'.

Théorème D.- Si V est une surface sans ombilic sur laquelle k_1 et k_2 sont partout indépendantes, il suffit que la surface $\tilde{V}$ ait avec V un G-contact global d'ordre 3 pour assurer la G-congruence locale.

Théorème E.- Si V est un plan ou une sphère, il suffit que la surface $\tilde{V}$ ait avec V un G-contact global d'ordre 2 pour assurer la G-congruence.

Par contre, dès que la sous-variété V n'est plus aussi régulière, le G-contact global de tous les ordres peut être insuffisant pour assurer la G-congruence.

Exemple.- $M = \mathbb{R}^2$, $G = E(2)$ groupe des déplacements. Soit $\mathcal{C}$ la courbe d'équation $y = f(x)$ où f est une fonction C^∞, nulle et infiniment plate en $x = 0$, décroissante sur $\mathbb{R}^-$ et croissante sur $\mathbb{R}^+$. Soit pour $a > 0$ la courbe $\tilde{C}_a$ d'équation $y = g_a(x)$ avec

$$g_a(x) = f(x) \qquad \text{pour } x < 0$$

$$g_a(x) = 0 \qquad \text{pour } x \quad [0,a]$$

$$g_a(x) = f(x - a) \qquad \text{pour } x > a$$

Il est clair que, pour tout $a > 0$ les courbes $\tilde{\mathcal{C}}_a$ et $\mathcal{C}$ ont un G-contact global de tous les ordres: pourtant elles ne sont pas égales, ni même localement G-congruentes au voisinage du point $x = 0$ de $\mathcal{C}$!.

Dans cet exemple, le G-contact global est trop "anarchique". Pour éviter ces situations, la notion plus fine suivante sera indispensable.

Définition 3.- On dit que deux sous-variétés V et $\tilde{V}$ ont un "G-contact local d'ordre k" si elles ont un G-contact global d'ordre k et, si de plus, pour toute couple $x \in V$, $\tilde{x} \in \tilde{V}$ où le G-contact (ponctuel) d'ordre k est realisé, on a la propriété suivante:

Pour toute voisinage ouvert U de x dans V, il existe un voisinage ouvert $\tilde{U}$ de $\tilde{x}$ dans $\tilde{V}$ tel que les sous-variétés U et $\tilde{U}$ aient un G-contact global d'ordre k et viceversa.

5.- Le Théorème de congruence classique.

a) Dans ce paragraphe $p < n$ est un entier fixé. La construction des invariants différentiels passe par celle des types à l'ordre k:

Un type à l'ordre k est une suite $\{W_i\}$, $o \leqslant i \leqslant k$. où chaque W_i est une sous-variété de $G_p^i(M)$ ayant les propriétés suivantes:

i) α_i induit une submersion de W_i sur W_{i-1} $(i = 1,2,...,k)$

ii) W_i rencontre au plus une fois chaque G-orbite et ceci transversalement au sens suivante:

$$(W_i)_x \cap T_x(G_x) = \{0\} \quad \forall\, x \in W_i$$

iii) W_i ne rencontre que des orbites de même dimension, ou ce qui est équivalent, la dimension $r_i(x)$ du groupe d'isotropie $G_i(x)$ du point x pour l'action de G est indépendante de $x \in W_i$.

Les types vérifient d'autres conditions techniques (cf. [8]).

b) <u>Définition 4.- La sous-variété $V = (S,f)$ est dite k-régulière du type de W_k si elle vérifie</u>

$$T^k f(S) \subset G \cdot W_k$$

Pour une telle variété on construit <u>l'indicatrice d'ordre k</u> qui est la fonction $f_{W_k} : S \longrightarrow W_k$ définie par

$$f_{W_k}(s) = (GT^k f(s)) \cap W_k$$

C'est une fonction différentiable et de plus, pour tout $s_0 \in S$, il existe un ouvert S_0, $s_0 \in S_0 \subset S$, et une application différentiable u de S_0 dans G telle que

$$f_{W_k}(s) = u(s)^{-1} T^k f(s) \qquad (5)$$

L'application u, qui n'est pas unique, est une <u>repère mobile d'ordre k de V</u> au sens d'Elie CARTAN (relativement à W_k ...) .

Si $\tilde{V} = (\tilde{S},\tilde{f})$ est aussi une sous-variété du type de W_k, alors V et $\tilde{V}$ ont un G-contact d'ordre k en $f(s_0)$ et $\tilde{f}(\tilde{s}_0)$, (resp. un G-contact global d'ordre k) si et seulement si $f_{W_k}(s_0) = \tilde{f}_{W_k}(\tilde{s}_0)$ (resp. $f_{W_k}(S) = \tilde{f}_{W_k}(\tilde{S})$). Si $V = (S,f)$ est du type de W_k, toute variété G-congruente

$\tilde{V} = g(V) = (S, L_g \circ f)$ est aussi du type de W_k et elles ont le même indicatrice.

Les fonctions indicatrices repèrent donc les orbites des éléments de contact d'ordre k pour les sous-variétés du type de W_k; elles fournissent les invariants différentielles d'ordre k de ces sous-variétés qui sont les fonctions $x^\alpha \circ f_{W_k}$ lorsque $\{x^\alpha\}$ est un système de coordonnées locales sur W_k.

Enfin, si V est du type de W_k, alors pour tout $i \leqslant k$ elle est i-régulière du type de W_i et l'on a

$$f_{W_{i-1}} = \alpha_i \circ f_{W_i} \qquad (1 \leqslant i \leqslant k)$$

Il découle de (5) qu'un repère mobile d'ordre k est aussi un repère mobile d'ordre i pour tout $i \leqslant k$.

c) Nous pouvons maintenant définir les repères de FRENET.

Définition 5.- Soit $\{W_0, W_1, \ldots, W_q\}$ un type à l'ordre q et V= (S,f) une sous-variété du type de W_q. On dit que q est "l'ordre des repères de Frenet" de V, (O.R.F.), si c'est le plus petit entier pour lequel

i) les orbites rencontrant W_{q-1} et W_q ont la même dimension $(r_{q-1} = r_q)$,

ii) les indicatrices $F_{W_{q-1}}$ et f_{W_q} sont des applications de même rang constant $(n_{q-1} = n_q)$.

On appelle alors "repère de FRENET" de V tout repère mobile d'ordre q.

Théorème 1.- (Théorème de congruence classique)

Soient $\{W_0, W_1, \ldots, W_q\}$ un type à l'ordre q et V une sous-variété du type de W_q dont q est l'ordre du repères de FRENET. Si une sous-variété $\tilde{V}$ a avec V un "G-contact local d'ordre q". alors V et $\tilde{V}$ sont "localement G-congruentes"; c'est-à-dire que si le G-contact ponctuel d'ordre q est realisé en $x \in V$ et $\tilde{x} \in \tilde{V}$. il existe un voisinage de x dans V et un voisinage de $\tilde{x}$ dans $\tilde{V}$ qui sont G-congruents.

Sous cette forme précise, le théorème est dû à S. HUCKEL. Dans le cas analytique il est dû à E. CARTAN, [4]; dans le cas C^∞ G.R. JENSEN, [7], donne un énoncé voisin. Les exemples du §2 sont tous des cas particulièrs de ce théorème.(Cf. aussi [5]).

Les limites et les inconvénients de ce théorème sont assez clairs: il y a l'arbitraire du choix des W_i pour construire les types, mais il y a surtout la restriction draconienne qu'impose à une sous-variété V le fait d'être q-régulière et de posséder des repères de FRENET. Ainsi, elle doit être k-régulière pour tout $k \leqslant q$ ce qui implique, en particulier, que pour tout $k \leqslant q$ les groupes d'isotropie de tous les éléments de contact d'ordre k de V doivent avoir la même dimension. Dans l'espace E^3 muni du group des déplacements, pour une courbe qui possède un point d'inflexion isolé x_0, — ou une surface qui possède en x_0 un ombilic isolé, — la dimension du group d'isotropie de l'élément de contact d'ordre 2 est 1 en x_0 alors qu'elle est nulle aux points voisins; ceci implique que, pour tout $k \geqslant 1$ une telle courbe ou surface n'est pas k-régulière et que le théorème ci-dessus ne peut donner des conditions suffisantes de congruence avec cette sous-variété au voisinage de x_0.

6.- Le Théoreme de Congruence de S. HUCKEL.

a) Définitions

Soient $V = (S,f)$ et $\tilde{V} = (\tilde{S},\tilde{f})$ deux sous-variétés de M. Les ensembles suivants s'introduisent naturellement:

i) L'ensemble $\Gamma_k(V,\tilde{V})$ des G-contacts d'ordre k de V et $\tilde{V}$.

$$\Gamma_k(V,\tilde{V}) = \{(s,\tilde{s},g) \in S \times \tilde{S} \times G \ / \ gT^k\tilde{f}(\tilde{s}) = T^k f(s)\}$$

on le munit de la topologie induite par $S \times S \times G$. Les inclusions suivantes découlent de (1):

$$\ldots \subset \Gamma_k(V,\tilde{V}) \subset \Gamma_{k-1}(V,\tilde{V}) \subset \ldots \subset \Gamma_1(V,\tilde{V}) \subset \Gamma_0(V,\tilde{V})$$

ii) L'ensemble $\Gamma_k(V) = \Gamma_k(V,V)$ des auto-contacts d'ordre k de V modulo G qui vérifient les inclusions

$$\{(s,s,e) \ / \ s \in S\} \subset \ldots \subset \Gamma_k(V) \subset \Gamma_{k-1}(V) \subset \ldots \subset \Gamma_1(V) \subset \Gamma_0(V)$$

On appelle ordre de stabilisation de V tout entier q tel que $\Gamma_q(V)$ soit un voisinage de $\{(s,s,e) \ / \ s \in S\}$ dans $\Gamma_{q-1}(V)$.

Nous aurons besoin d'une autre notion plus forte que le G-contact local.

Définition 6.- On dit que deux sous-variétés V et $\tilde{V}$ de M ont un "G-contact continu d'ordre k" si les projections

$$p_1 : (S \times \tilde{S} \times G) \longrightarrow S \qquad p_2 : (S \times \tilde{S} \times G) \longrightarrow \tilde{S}$$

restreintes à $\Gamma_k(V,\tilde{V})$ sont des applications ouvertes et surjectives.

Enfin, pour donner une condition de transversalité de $T^k f$ aux orbites de G dans $G_p^k(M)$ on est amené à considerer le sous-espace vectoriel de $T_x\, G_f(M)$.

$$\textstyle\sum_k(s) = (T^k f)_s^T(S_s) + T_{x_k}(G_{x_k}), \quad x_k = T^k f(s), \ s \in S.$$

Notons la propriété

$$\textstyle\sum_k(s) = \alpha_{k+1}^T \sum_{k+1}(s)$$

qui implique, entre autres, que la dimension de $\sum_k(s)$ croît avec k.

b) Avec ces définitions, on peut énoncer le théorème de congruence

Théorème 2. (S. HUCKEL [8]). Soient $V = (S,f)$ une sous-variété de M, q un ordre de stabilisation de V tel que la dimension de $\sum_{q-1}(s)$ soit indépendante de $s \in S$ (condition de transversalité). Si une sous-variété $\tilde{V}$ a avec V un "G-contact continu d'ordre q", alors V et $\tilde{V}$ sont localement G-congruentes.

Notons que ce théorème ne suppose pas que l'action de G sur M soit transitive; pas plus que le théorème 1 dans l'énoncé de S. HUCKEL, à la différence des énoncés antérieurs. En effet le théorème 1 se déduit du théorème 2 en analysant les propriétés des sous-variétés k-régulières. On montre en particulier, [8], que pour une variété k-régulière V

i) l'ordre des repères de FRENET de V est le plus petit entier $q \leq k$ tel que $\Gamma_q(V)$ et $\Gamma_{q-1}(V)$ soient tous deux des sous-variétés de $S \times \tilde{S} \times G$ et de la même dimension $2p - n_q + r_q$, de sorte que l'O.R.F. q de V est un ordre de stabilisation de V. Notons au passage que ceci constitue une définition intrinsèque de l'O.R.F., indépendante de la construction particulière du type $\{W_0, W_1, \ldots, W_q\}$.

ii) Si $q \leq k$ est l'O.R.F. de V, la dimension de $\sum_{q-1}(s)$ est indépendante de $s \in S$, et si $\tilde{V}$ a un G-contact local d'ordre q avec V, alors V et $\tilde{V}$ ont un G-contact continu d'ordre q.

7.- Exemples d'applications du théorème 2.

Revenons à l'espace euclidean E^3

a) Le théorème 2 permet de classifier à déplacement prés les courbes qui ont un point d'inflexion isolé non infiniment plat. Soit $s \longrightarrow \overrightarrow{f(s)}$ une représentation normale d'une telle courbe C et $\overrightarrow{t(s)} = \overrightarrow{f'(s)}$ le vecteur tangent unitaire. Si le point d'inflexion a pour abscisse curviligne $s = 0$, notons m le plus petit entier $\geqslant 3$ tel que la dérivée $\overrightarrow{f^{(m)}}(0)$ soit non nulle. Soit alors $(\overrightarrow{t(s)}, \overrightarrow{\nu(s)}, \overrightarrow{\beta(s)})$ la base orthonormée positive de E^3 définie, pour s voisin de zéro par

$$\overrightarrow{f^{(m)}}(s) = \lambda(s)\overrightarrow{t(s)} + \mu(s)\overrightarrow{\nu(s)} \qquad \mu(s) > 0$$

et $b = \frac{\overrightarrow{dt}}{ds} \cdot \overrightarrow{\beta}$. S. HUCKEL a établi à l'aide du théorème 2 qu'une courbe C' ayant avec C un G-contact continu d'ordre k lui est localement G-congruente si k a les valeurs suivantes:

$$k = m + 1 \quad \begin{cases} \text{si } m \text{ est impair} \\ \quad \text{ou} \\ \text{si } m \text{ est pair, la première dérivée de } b \text{ non nulle} \\ \text{en zéro étant d'ordre impair} \end{cases}$$

$$k = m + 2 \qquad \text{si } m \text{ est pair, dans les autres cas.}$$

b) Surfaces ayant un ombilic isolé.

Pour une surface S de E^3, on a pour tout $k \in \mathbb{N}$

$$\dim \textstyle\sum_k (s) \leqslant 2 + \dim G = 8$$

Soit S une surface ayant en 0 un ombilic isolé, tangente en ce point au plan Oxy, où $Oxyz$ un repère orthonormé. Dans ce repère S a une équation locale $z = \phi(x,y)$, ϕ fonction C^∞ au voisinage de $(0,0)$. On voit que $\dim \sum_2(s) = 8$ pour $s \neq 0$ ce qui implique

$$\dim \textstyle\sum_k(s) = 8 \text{ pour } s \neq 0, \qquad \geqslant 2$$

$$\dim \textstyle\sum_2(0) = 5 + \operatorname{rang}\begin{pmatrix} \phi_{x^3} & , & \phi_{x^2y} \\ \phi_{x^2y} & , & \phi_{xy^2} \\ \phi_{xy^2} & , & \phi_{y^3} \end{pmatrix}(0) \leqslant 7$$

La condition de transversalité du théorème 2 n'est donc jamais vérifiée pour $q-1 = 2$; il faudra $q \geqslant 4$ pour appliquer le théorème 2 au voisinage de l'ombilic.

<u>Cas des surfaces de révolution</u> autour de l'axe Oz : les dérivées partielles d'ordre 3 de ϕ en zero sont nulles et l'on a

$$\dim \Sigma_2(0) = \dim \Sigma_3(0) = 5$$

mais si le développement de Taylor de ϕ en (0,0) est

$$\phi(x,y) = \frac{x^2 + y^2}{2p} + A(x^2 + y^2) + (\text{ ordre} > 4) \quad \ldots \quad \text{avec } A \neq 0,$$

alors $\dim \Sigma_4(0) = 8 = \dim \Sigma_4(s)$ $(s \neq 0)$. La condition de transversalité est vérifiée pour $q - 1 = 4$ et comme d'autre part

$$\Gamma_k(S) = \Gamma_3(S)$$

pour $k \geqslant 3$, le théorème 2 s'applique déjà pour $q = 5$: <u>le G-contact continu d'ordre 5 d'une surface $\tilde{S}$ avec S assure la G-congruence au voisinage de l'ombilic</u>.

<u>Cas d'une surface ayant un plan de symétrie</u>.- Si ce plan est le plan Oyz , le développement de Taylor en (0,0) de ϕ s'écrit

$$\phi(x,y) = \frac{x^2 + y^2}{2p} + ax^2y + by^3 + (\text{ordre} > 3) \ldots$$

Pour $ab \neq 0$, on voit que

$$\dim \Sigma_3(0) = 8 = \dim \Sigma_3(s) \ , \ s \neq 0 \ .$$

de sorte que la condition de transversalité est vérifiée pour $q \geqslant 4$. Comme, sous des hypothèses suffisamment générales,

$$\Gamma_k(S) = \Gamma_3(S) = \{(s,s,e) \ / \ s \in S\}$$

pour $k \geqslant 3$, le théorème 2 s'applique pour $q = 4$: le <u>G-contact continu d'ordre 4 assure la G-congruence au voissinage de l'origine</u>.

Dans ce trois exemples, on a pu appliquer le théorème 2 en "effaçant la singularité à l'ordre 2 (inflexion,ombilic) par le passage à un ordre supérieur convenable, procédure qui est essentiellement impossible avec le théorème 1.

BIBLIOGRAPHIE

[1] W. Ambrose. Higher order Grassmann bundles. Topology. Vol 3. suppl. 2. 199-238. (1965)

[2] D. Bernard. Espaces homogènes et repère mobile. IV. Coll. Int. de Geom. Diff. Santiago de Compostela 55 - 63 (1978).

[3] D. Bernard. Immersions et repères mobiles. Symposium E.B. Christoffel. Birkhäuser. Basel. (1981).

[4] E. Cartan. La théorie des groupes finis et continus et la géométrie différentielle par la méthode du repère mobile. Gauthier-Villars. Paris. 1937.

[5] M.L. Green. The moving frame, differential invariants and rigidity theorems for curves in homogeneous spaces. Duke Math. J. 45, 735-779 , (1978).

[6] P. Griffiths. On Cartan's method of Lie Groups and moving frames as applied to uniqueness and existence questions in differential geometry. Duke. Math. J. 41, 775-814 , (1974).

[7] G.R. Jensen. Higher order contact of submanifolds of homogeneous spaces. Lecture Notes in Mathematics 610. Springer. Berlin, Heidelberg, (1977).

[8] S. Huckel. Invariants différentiels, G-contact. théorèmes de congruence. Thèse. Strasbourg. (1981).

KILLING VECTOR FIELDS AND COMPLEX STRUCTURES

C. Currás-Bosch
Dpto. Geometría y Topología
Fac. Matemáticas, Universidad Barcelona
SPAIN

Let (M,g) be a complete connected Riemannian manifold of class C^{∞} and let X be a Killing vector field (K.v.f.). It is well known (see for instance [4]) that the $(1,1)$ operator $A_X = L_X - \nabla_X = -\nabla X$ (L represents the Lie derivation and ∇ the derivation with respect to the Levi-civita connection) is skew-symmetric.

At each point $x \in M$ we can decompose the Lie algebra of skew-symmetric endomorphisms of $T_x(M)$, $E(x)$, as $G(x) + G(x)^{\perp}$, where $G(x)$ is the holonomy algebra at x and $G(x)^{\perp}$ the orthogonal complement with respect to the Killing form $\langle A,B\rangle = -(1/\dim.M)\,\mathrm{trace}\,(A\circ B)$.

For each Killing vector field we can decompose $A_X = S_X + B_X$, $(S_X)_x$ lying in $G(x)$ at each $x \in M$ and $B(_X)_x$ in $G(x)^{\perp}$. It is well known that B_X is parallel, because for each $Y \in T(M)$, $\nabla_Y(A_X) = R(X,Y) = \nabla_Y(S_X) + \nabla_Y(B_X)$ (see [4]), and $\nabla_Y(B_X)$ is the only term lying in $G(x)^{\perp}$.

We say that X is a non holonomic Killing vector field (n.h.K.v.f.) if and only if $B_X \neq 0$.

Kostant,[5],proved that if M is compact then $B_X = 0$ for each K.v.f. X ; the technical arguments used by Kostant can not be applied to study the non compact case; it is necessary to use other methods; in this order we stated in [3] a sufficient condition to ensure that in a complete and non compact Riemannian manifold, a K.v.f.X verifies $B_X = 0$.

In this lecture we are going to state necessary conditions over complete Riemannian manifolds for them to admit n.h.K.v.f.'s.

By using the universal covering of those manifolds we reduce this problem to irreducible Riemannian ones. We associate to the set of n.h.K.v.f.'s a Lie algebra and we prove that this Lie algebra can be either $\mathbb{R}$ or so(3) . We also prove that the manifold must be either Kähler with Ricci tensor null or hyperKähler.

Finally by using some examples of these manifolds, given by Calabi [1], we find some n.h.K.v.f.'s over irreducible manifolds.

1.- First results

Throughout this lecture, let (M,g) be a complete connected Riemannian manifold of class C^∞. Let X be a K.v.f. and let $\tilde{M}$ be the universal covering of M, endowed with the induced Riemannian metric. Let $\tilde{X}$ be the induced K.v.f. from X; then it is well known that $A_{\tilde{X}} = A_X$ at corresponding points..

Because de Rham's decomposition theorem $\tilde{M} = M_0 \times M_1 \times \ldots \times M_k$, where M_0 is an euclidean space and each M_i $(1 \leqslant i \leqslant k)$ is irreducible, then $\tilde{X}$ and $A_{\tilde{X}}$ reads as follows : $\tilde{X} = X_0 + X_1 + \ldots + X_k$, $A_{\tilde{X}} =$ $= A_{X_0} + A_{X_1} + \ldots + A_{X_k}$, where each A_{X_α} $(0 \leqslant \alpha \leqslant k)$ acts only on $T(M_\alpha)$.

Obviously B_X will be non zero if and only if $B_{X_\alpha} \neq 0$ for some α. If $A_{X_0} \neq 0$, as M_0 is euclidean $A_{X_0} = B_{X_0} \neq 0$, so if X_0 is not parallel then $B_{X_0} \neq 0$ so that $B_X \neq 0$ and X is a n.h.K.v.f. .

Because of the above result we are going to study from now on, under which conditions an irreducible Riemannian manifold can admit some n.h.K.v.f. . Obviously we can omit the simply connectedness assumption.

First of all if we suppose that there exists some n.h.K.v.f.,X in such a manifold, from the irreducible character and the parallelism of B_X we deduce that $(B_X)_x$ is an automorphism of $T_x(M)$ for each $x \in M$ and $(B_X)^2 = -k\mathrm{Id}.$, $k \in \mathbb{R}^+$. Taking the K.v.f. $(1/\sqrt{k})X$, it follows that $(B_{(1/\sqrt{k})X})^2 = -\mathrm{Id}.$, so that $B_{(1/\sqrt{k})X}$ is an almost complex structure.

From now on we can suppose $k = 1$ and as $\nabla(B_X) = 0$, $(M,B_X;g)$ is a Kähler manifold and as $(B_X)_x \in G(x)^\perp$ for each $x \in M$, we deduce from

$S(U,V) = (1/2)\{\text{trace}\, B_X \circ R(U, B_X(V))\}$, (S = Ricci tensor), that the Ricci tensor is null and so the holonomy algebra must be contained in $su(n)$ ($\dim_{\mathbb{R}}(M) = 2n$).

Now we prove some technical results which will be useful later.

Lemma 1.1.- Let S be a (1,1) tensor lying in $G(x)$ at each point $x \in M$; then $(L_X S)_x \in G(x)$, for all $x \in M$.

Proof.- $L_X S = L_X S - \nabla_X S + \nabla_X S = [A_X, S] + \nabla_X S$, but $(\nabla_X S)_x \in G(x)$ and $[A_X, S]_x \in G(x)$ because as it is well known $(A_X)_x$ belongs to the normalizer of $G(x)$.

Lemma 1.2.- Let B be a (1,1) tensor lying in $G(x)^{\perp}$ at each point $x \in M$; then $(L_X B)_x \in G(x)^{\perp}$, for all $x \in M$.

Proof.- We take any S in G and then we have $L_X(-\text{trace}(S \circ B)) = 0 =$ $= -\text{trace}\,(L_X S \circ B) - \text{trace}\,(S \circ L_X B)$, and as $L_X S \in G$, we get $L_X B \in G^{\perp}$.

Lemma 1.3.- Let X, Y be K.v.f.'s then we have,

$$[B_X, B_Y] = B_{[X,Y]} = L_X(B_Y) = -L_Y(B_X) \ .$$

Proof.- $L_X(B_Y) = L_X(B_Y) - \nabla_X(B_Y) = [A_X, B_Y] = [S_X, B_Y] + [B_X, B_Y] = [B_X, B_Y]$, because B_Y is the centralizer of G .

As $[A_Y, A_X] = -A_{[X,Y]} + R(X,Y)$ (see [4]), we have :

$$[S_Y, S_X] + [B_Y, B_X] = -S_{[X,Y]} - B_{[X,Y]} + R(X,Y)$$

so $[B_Y, B_X] = -B_{[X,Y]}$.

Corollary 1.1.- Let X, Y be K.v.f.'s, then we have,

$$[S_X, S_Y] - S_{[X,Y]} + R(X,Y) = 0$$

(Observe that this result remains valid without the irreducibility assumption).

Now we know some necessary conditions which must be accomplished by M if there exists some n.h.K.v.f. . To obtain more information we are going to study the set of n.h.K.v.f.'s.

2.- Main result

Let $i(M)$ be the Lie algebra of Killing vector fields and we consider for each $x \in M$ the map,

$$\begin{aligned} \phi : i(M) &\longrightarrow E(x) \\ X &\longrightarrow (B_X)_x \end{aligned}$$

From the parallelism of B_X we see that the point x under consideration can be anyone and from Lemma 1.3 ϕ is a Lie algebras morphism.

As the kernel of this morphism is the complementary in $i(M)$ of the set of n.h.K.v.f.'s , we see that the algebraic object associated in a natural way to the set of n.h.K.v.f.'s is the Lie algebra $\phi(i(M)) = L \subset E(x)$.

Let us suppose that dim.$L > 1$, then take $B, C \in L$. As for each $D \in L$, $D^2 = -\langle D,D\rangle Id.$, being $\langle D,D\rangle = -(1/\dim.M)\,\mathrm{trace}\, D^2$, from $(B + C)(B + C) = B^2 + C^2 + (BC + CB)$, we obtain,

$-\langle B + C, B + C\rangle Id. = -\langle B,B\rangle Id. - \langle C,C\rangle Id. + (BC + CB)$, so $(BC + CB) \equiv -2\langle B,C\rangle Id.$. Taking $B \perp C$, it follows that $(BC + CB) = 0$.

So B and C **span** a Lie subalgebra of $L, L' = \{B, C, BC = (1/2)[B,C]\} = so(3)$.

Suppose now that $L' \subsetneq L$, then there exists $E \in L$ and $E \perp L'$; by the same argument as above we obtain, $EB = -BE$, $EC = -CE$, and $E(BC) = -(BC)E$ which is wrong.

So L can be either $\mathbb{R}$ or $so(3)$.

If $L = so(3)$, we have on M two complex structures I, J associated in the above form to two n.h.K.v.f.'s, such that with respect to them (M,g) is Kähler and $IJ = -JI$. In this case it is well known that M is endowed with a quaternionic structure, also called, see Calabi [1], hyperKähler structure. In this case $\dim.M = 4k$, and the holonomy algebra is contained in $sp(k)$.

So we have obtained,

Theorem 2.1.- Let (M,g) be a complete, connected, irreducible, Riemannian manifold such that $L \neq 0$ (with some n.h.K.v.f.). Then

a) If $\dim .M = 2(2k+1)$, $L = \mathbb{R}$ and M is a Kähler manifold such that the holonomy algebra is contained in $su(2k+1)$.

b) If $\dim .M = 4k$, we have either i) $L = \mathbb{R}$ and M verifies the same conditions that in a) or ii) $L = so(3)$ and M is endowed with a hyperKähler structure.

3.- Examples

Finally we would like to give some examples of n.h.K.v.f.'s over irreducible Riemannian manifolds.

In order to do so, we know from Theorem 2.1, that M must be either Kähler or hyperKähler, and we are going to use some examples of these structures given by Calabi in [1] and [2].

First example

We consider an irreducible, non compact, Kähler manifold with null Ricci tensor (see [1]) pag. 284-285). One takes the canonical line bundle on $P_n(\mathbb{C})$ and the local Kähler potential defined on $\pi^{-1}(U)$, being U a subdomain of inhomogeneous coordinate functions $(z^1,\ldots,z^n)$, by

$$\Psi = \Phi \circ \pi + u(t)$$

Φ is the local Kähler potential on U for the Fubini-Study metric

$$\Phi = (1/k)(1 + k(\sum_{i=1}^{n} |z^i|^2)$$

$$t = \exp((n+1)/2)k\Phi)|\xi|^2 \qquad (\xi \text{ is the fibre coordinate})$$

$$u(x) = u_o + (2n/n+1)(\sqrt[n]{1+cx} - 1) -$$

$$- \sum_{j=1}^{n-1} (2(1-\omega^j)/(n+1))\log((\sqrt[n]{1+cx} - \omega^j/(1-\omega^j))$$

$(\omega = \exp(2\pi i/n))$

One can prove, (see [1]), that this manifold is an irreducible, complete, non compact, Kähler manifold such that its holonomy algebra is contained in $su(n+1)$ (Ricci tensor null).

As Ψ depends only on $|z^i|^2$ and $|\xi|^2$, it is easy to see that the vector field $X = i\xi(\partial/\partial\xi) - i\bar{\xi}(\partial/\partial\bar{\xi})$ is an infinitesimal automorphism and as $X(\Psi) = 0$, is also a Killing vector field.

A simple calculation shows that the (1,1) operator A_X does not lie in $su(n+1)$ at the points where $\xi = 0$.

The K.v.f. X defined in $\pi^{-1}(U)$ can be extended by considering similar expressions in the other subdomains of inhomogeneous coordinates and we obtain a K.v.f. such that $B_X \neq 0$ at some points, so $B_X \neq 0$ everywhere.

Furthermore as the holonomy algebra of these manifolds is $su(n+1)$ (for $n > 1$), we have $L = \mathbb{R}$, because if $L = so(3)$ the holonomy algebra must be contained in $sp(\nu)$ ($2\nu = n+1$, if it is possible), which for $\nu > 1$, is strictly contained in $su(2\nu)$.

Second example

We consider the example of hyperKähler manifolds furnished by Calabi in [1] and [2], to find just in the same form as above some n.h.K.v.f.'s.

Let us take the holomorphic cotangent bundle to $P_n(\mathbb{C})$, $T^{*\prime}(P_n(\mathbb{C}))$ and take on $\pi^{-1}(U)$, being U a subdomain of inhomogeneous coordinates $(z^1,\dots,z^n)$ and $(\xi_1,\dots,\xi_n)$ the fibre cooordinates, the local Kähler potential given by,

$$\Psi(z^i;\xi_j) = \log(1 + |z|^2) + \sqrt{1+4t} - \log(1 + \sqrt{1+4t});$$

where $t = (1+|z|^2)(|\xi|^2 + |z\xi|^2)$; $|z|^2 = \sum_{i=1}^{n} |z^i|^2$, $|\xi|^2 = \sum_{j=1}^{n} |\xi_j|^2$; $z\xi = \sum_{i=1}^{n} z^i\xi_i$.

Let us consider also the (2,0) 2-form $H = \sum_{i=1}^{n} dz^i \wedge d\xi_i$.

Calabi proved that Ψ is the Kähler potential for a metric g which is hyperKähler with respect to the given complex structure and H.

By the same argument as above one can see that the vector field

$X = \sum_{j=1}^{n} iz^j(\partial/\partial z^j) - i\bar{z}^j(\partial/\partial \bar{z}^j)$ is a Killing vector field (defined on $\pi^{-1}(U)$, such that $L_X H \neq 0$, so X is not an infinitesimal automorphism with respect to the complex structure associated to H via g ; so $B_X \neq 0$.

The globalization of the K.v.f. X, can be made by using similar expressions on the remaining subdomains of inhomogeneous coordinates.

All the examples of n.h.K.v.f.'s, that we have obtained in this case, by using similar expressions, for instance $\sum_{j=1}^{n} i\xi_j(\partial/\partial\xi_j) - i\bar{\xi}_j(\partial/\partial\bar{\xi}_j)$, verify that its B_- part is always the same, unless a constant, so we do not know if it is possible, in this case, to find another n.h.K.v.f. furnishing with the above one, by considering their B_- parts, the hyperKähler structure.

REFERENCES

[1] Calabi,E. "Métriques Kählériennes et fibrés holomorphes" Ann. Sci. Ec.Norm.sup. 4e série, 12, 1979, 269-294.

[2] Calabi,E. "Isometric families of Kähler structures" The Chern Symposium (1979), Springer 1980.

[3] Currás-Bosch,C. "Campos de Killing y álgebras de holonomía" Actualités mathématiques, Actes du VI C.G.M.E.L., Gauthier-Villars, Paris 1982.

[4] Kobayashi,S.-Nomizu,K. "Foundations of Differential Geometry" Vol.-I, Interscience Publ.,1963.

[5] Kostant,B. "Holonomy and the Lie algebra of infinitesimal motions of a Riemannian manifold". Trans. Amer. Math. Soc. 80, (1955).

[6] Lichnérowicz,A. "Géométrie des groupes de transformations". Dunod, Paris 1958.

DERIVATIONS IN THE TANGENT BUNDLE

J.J. Etayo
Facultad de Matemáticas
Universidad Complutense
Madrid (Spain)

1.- The aim of this paper is the systematization of some results of [2] and [3] .

Let us recall that in a differentiable manifold M , a derivative in the direction of a vector field X , D_X , is determined locally by the components of the vector field X^i , and the functions Q^i_j defined by

$$D_X\left(\frac{\partial}{\partial x^j}\right) = \left(\frac{\partial}{\partial x^i}\right)Q^i_j \quad .$$

Then the components of D_X with respect to the local system $(x^1,\ldots,x^n)$ are $(X^i;Q^i_j)$ ([4]).

A derivation D is an $\mathbb{R}$-linear map from the module of vector fields, $T^1_o(M)$, to the module of the directional derivatives considered before. The image under D of a vector field X is called D_X . Moreover if D is $T^o_o(M)$-linear the derivation becomes a connection. The local components of a derivation D are (dx^i,ϕ^i_j) where $\phi^i_j : T^1_o(M) \longrightarrow T^o_o(M)$ are $\mathbb{R}$-linear mappings such that $\phi^i_j(X) = Q^i_j$. If ϕ^i_j are $T^o_o(M)$-linear, D will be a connection.

We consider the tangent bundle $\pi : TM \longrightarrow M$. Each chart $(x^1,\ldots,x^n)$ of M induces one of TM $(x^1,\ldots,x^n,x^{n+1},\ldots,x^{2n})$ and a tangent vector at p , X_p , has coordinates $(p^1,\ldots,p^n,a^1,\ldots a^n)$ where $p^i = x^i(p)$, $X_p = \left(\frac{\partial}{\partial x^i}\right)_p a^i$.

For each function $f \in T^o_o(M)$, we consider the function $f^v = f\circ\pi \in T^o_o(TM)$ and we call it the <u>vertical lift</u> of f . Also, we define the function

$$f^c = \frac{\partial f}{\partial x^i} x^{n+i} \in T^o_o(TM)$$

that is called the complete lift of f.
A vector field of TM , $\bar{X}$, is said to be projectable if for any $f \in T^o_o(M)$, $\bar{X}f^v \in T^o_o(M)$. If $\begin{pmatrix}\bar{X}^i \\ \bar{X}^{n+i}\end{pmatrix}$ are the coordinates of $\bar{X}$, it is projectable if and only if $\bar{X}^i \in T^o_o(M)$. In this case, the vector field $X = \frac{\partial}{\partial x^i} \bar{X}^i \in T^1_o(M)$ is called the projection of $\bar{X}$ and we write $X = \pi_*(\bar{X})$. The set of projectable vector fields is a $T^o_o(M)$-module but not a $T^o_o(TM)$-module. A vector field in TM is vertical if and only if it projects to the null vector field. They form a $T^o_o(TM)$-module that we denote $V(TM)$.

Now, let Y be a vector field in M with coordinates (Y^i). We call the complete lift of Y , $Y^c \in T^1_o(TM)$ the vector field of coordinates $\begin{pmatrix} Y^i \\ (Y^i)^c \end{pmatrix}$.

Also the vertical lift Y^v is the vertical vector field of coordinates $\begin{pmatrix} 0 \\ Y^i \end{pmatrix}$.

For a detailed study of those lifts see [4].

In [2], we defined the vertical and complete lifts of directional derivatives as follows : if (X^i, Q^i_j) are the components of D_X

$$v(D_X) = \left(\begin{array}{c|cc} 0 & 0 & 0 \\ X^i & Q^i_j & 0 \end{array}\right)$$

and

$$C(D_X) = \left(\begin{array}{c|cc} X^i & Q^i_j & 0 \\ (X^i)^c & (Q^i_j)^c & Q^i_j \end{array}\right)$$

are the components of the vertical and the complete lifts respectively.

In short this is denoted by

$$v(D_X) = (X^V, Q^V)$$

and

$$c(D_X) = (X^C, Q^C) \ .$$

2.- Now let $D : (dx^i ; \phi^i_j)$ be a derivation of M ; our purpose is to define

$$D^C \ : \ \left\{ \begin{array}{l|ll} dx^i & \bar{\phi}^i_j & \bar{\phi}^i_{n+j} \\ dx^{n+i} & \bar{\phi}^{n+i}_j & \bar{\phi}^{n+i}_{n+j} \end{array} \right\}$$

where each $\bar{\phi}$ is a $\mathbb{R}$-linear mapping of $T^1_o(TM)$ into $T^o_o(TM)$, in such a way that we can call it the <u>complete lift</u> of D . It seems natural to impose the condition

$$D^C_{X^C} Y^C = (D_X Y)^C$$

for any vector fields $X, Y \in T^1_o(M)$

If $X : (X^i)$, $Y : (Y^i)$, we obtain :

$$D_X Y = D_X\left\{\frac{\partial}{\partial x^h} \cdot Y^h\right\} = \frac{\partial}{\partial x^h} \cdot X(Y^h) + D_X\left\{\frac{\partial}{\partial x^h}\right\} \cdot Y^h =$$

$$= \frac{\partial}{\partial x^h} \cdot \frac{\partial Y^h}{\partial x^j} \cdot X^j + \frac{\partial}{\partial x^k} \cdot \phi^k_h(X) Y^h = \frac{\partial}{\partial x^k} \cdot \left\{\frac{\partial Y^k}{\partial x^j} \cdot X^j + \phi^k_h(X) Y^h\right\} .$$

Then :

$$(D_X Y)^C = \frac{\partial}{\partial x^k} \cdot \left\{\frac{\partial Y^k}{\partial x^j} \cdot X^j + \phi^k_h(X) Y^h\right\} + \frac{\partial}{\partial x^{n+k}} \cdot \left\{\frac{\partial Y^k}{\partial x^j} X^j + \phi^k_h(X) Y^h\right\}^C \ .$$

Since

$$X^C = \frac{\partial}{\partial x^i} \cdot X^i + \frac{\partial}{\partial x^{n+i}} \cdot (X^i)^C \quad ,$$

and the same Y^C , we have

$$D^c_{X^c}Y^c = D^c_{X^c}\left(\frac{\partial}{\partial x^h}\cdot Y^h + \frac{\partial}{\partial x^{n+h}}\cdot (Y^h)^c\right) =$$

$$= \frac{\partial}{\partial x^h}\cdot X^c(Y^h) + D^c_{X^c}\left(\frac{\partial}{\partial x^h}\right)\cdot Y^h + \frac{\partial}{\partial x^{n+h}}\cdot X^c(Y^h)^c + D^c_{X^c}\left(\frac{\partial}{\partial x^{n+h}}\right)\cdot (Y^h)^c$$

$$= \frac{\partial}{\partial x^h}\cdot X^c(Y^h) + \left(\frac{\partial}{\partial x^k}\cdot \bar{\phi}^k_h(X^c) + \frac{\partial}{\partial x^{n+k}}\, \bar{\phi}^{n+k}_h(X^c)\right)Y^h +$$

$$+ \frac{\partial}{\partial x^{n+h}}\cdot X^c(Y^h)^c + \left(\frac{\partial}{\partial x^k}\cdot \bar{\phi}^k_{n+h}(X^c) + \frac{\partial}{\partial x^{n+k}}\cdot \bar{\phi}^{n+k}_{n+h}(X^c)\right)(Y^h)^c$$

$$= \frac{\partial}{\partial x^k}\left(X^c(Y^k) + \bar{\phi}^k_h(X^c)\cdot Y^h + \bar{\phi}^k_{n+h}(X^c)\cdot (Y^h)^c\right) +$$

$$+ \frac{\partial}{\partial x^{n+k}}\left(X^c(Y^h)^c + \bar{\phi}^{n+k}_h(X^c)\cdot Y^h + \bar{\phi}^{n+k}_{n+h}(X^c)\cdot (Y^h)^c\right).$$

By equalizing :

$$\frac{\partial Y^k}{\partial x^j}X^j + \phi^k_h(X)Y^h = X^c(Y^k) + \bar{\phi}^k_h(X^c)Y^h + \bar{\phi}^k_{n+h}(X^c)(Y^h)^c$$

$$\frac{\partial Y^k}{\partial x^j}(X^j)^c + \left(\frac{\partial Y^k}{\partial x^j}\right)^c X^j + \phi^k_h(X)(Y^h)^c + (\phi^k_h(X))^c Y^h =$$

$$= X^c(Y^h)^c + \bar{\phi}^{n+k}_h(X^c)Y^h + \bar{\phi}^{n+k}_{n+h}(X^c)(Y^h)^c \quad .$$

Because

$$X^c(Y^k) = \frac{\partial Y^k}{\partial x^j}\cdot X^j + \frac{\partial Y^k}{\partial x^{n+j}}(X^j)^c = \frac{\partial Y^k}{\partial x^j}X^j \ ,$$

$$X^c(Y^k)^c = \frac{\partial (Y^k)^c}{\partial x^j}X^j + \frac{\partial (Y^k)^c}{\partial x^{n+j}}(X^j)^c = \left(\frac{\partial Y^k}{\partial x^j}\right)^c X^j + \frac{\partial Y^k}{\partial x^j}(X^j)^c \ ,$$

It results :

$$\bar{\phi}^k_h(X^c) = \phi^k_h(X) \ , \qquad \bar{\phi}^k_{n+h}(X^c) = 0$$

$$\bar{\phi}^{n+k}_h(X^c) = (\phi^k_h(X))^c \ , \qquad \bar{\phi}^{n+k}_{n+h}(X^c) = \phi^k_h(X) \ ,$$

relations which will be verified by the components of the derivation D^c lifted to the tangent bundle.

3.- The above relations and the definitions of the lifts of the vector fields suggest the following definitions of lifts of maps $\phi \in \mathrm{Hom}_{\mathbb{R}}(T^1_o(M), T^o_o(M))$.

Let $\bar{\phi} \in \mathrm{Hom}_{\mathbb{R}}(T^1_o(TM), T^o_o(TM))$, $\bar{\phi}$ is projectable if for any projectable field of TM , $\bar{X}$, $\bar{\phi}(\bar{X})$ lies in $T^o_o(M)$. If $X = \pi_*(\bar{X})$ we define $\phi(X) = \bar{\phi}(\bar{X})$. Obviously ϕ is well defined and $\phi \in \mathrm{Hom}_{\mathbb{R}}(T^1_o(M), T^o_o(M))$. It is called the projection of $\bar{\phi}$ and it is determined by

$$\phi(X) = \bar{\phi}(X^C) .$$

As $\bar{\phi}(X^V) = 0$, we are in a similar situation to the tensor field and we can define the lifts.

Let $\phi \in \mathrm{Hom}_{\mathbb{R}}(T^1_o(M), T^o_o(M))$. $\phi^V \in \mathrm{Hom}_{\mathbb{R}}(T^1_o(TM), T^o_o(TM))$ is a <u>vertical lift</u> of ϕ if it satisfies $\phi^V(X^C) = \phi(X) = (\phi(X))^V$; $\phi^V(X^V) = 0$.

In this case, ϕ^V is projectable and its projection is ϕ . Notice that the above conditions uniquely determine ϕ^V on the projectable vector fields but not over all fields of TM. In the most important particular cases we can choose a unique lifting by adding some extra condition.

Similarly, a map $\phi^C \in \mathrm{Hom}_{\mathbb{R}}(T^1_o(TM), T^o_o(TM))$ is called a <u>complete lift</u> of ϕ if it satisfies $\phi^C(X^C) = (\phi(X))^C$; $\phi^C(X^V) = (\phi(X))^V = \phi(X)$.

Thereby, a complete lift of the derivation $D : (dx^i; \phi^i_j)$ is given locally by

$$D^C : \left(\begin{array}{c|cc} (dx^i)^V & (\phi^i_j)^V & \bar{\phi}^i_{n+j} \\ (dx^i)^C & (\phi^i_j)^C & (\phi^i_j)^V \end{array}\right)$$

with $\bar{\phi}^i_{n+j}(X^C) = 0$.

We shall study several cases where $\bar{\phi}^i_{n+j}$ can be chosen in a canonical way.

4.- First we consider ϕ a linear differential form, i.e. $\phi = \omega \in T^o_1(M)$. We can chose both the vertical and complete liftings to be linear forms of TM and this determines them, ([4]). If (ω_i) are the components of ω , then $\omega^V = (\omega_i, 0)$; $\omega^C = ((\omega_i)^C, \omega_i)$.

If $D : (dx^i; \omega^i_j)$ with $\omega^i_j \in T^0_1(M)$, D is a connection that we call ∇ and ω^i_j are the connection forms. We choose canonically the complet lift of ∇, ∇^C , such that

$$\nabla^C \quad : \quad \left(\begin{array}{c|cc} (dx^i)^V & (\omega^i_j)^V & \bar{\phi}^i_{n+j} \\ (dx^i)^C & (\omega^i_j)^C & (\omega^i_j)^V \end{array}\right)$$

where $\bar{\phi}^i_{n+j}$ are differential forms satisfying $\phi^i_{n+j}(X^C) = 0$. But this condition implies $\bar{\phi}^i_{n+j} = 0$, so the complete lift of the connection ∇ is the connection

$$\nabla^C \quad : \quad \left(\begin{array}{c|cc} (dx^i)^V & (\omega^i_j)^V & 0 \\ (dx^i)^C & (\omega^i_j)^C & (\omega^i_j)^V \end{array}\right)$$

This result was also obtained in [2] and [4] by different methods.

5.- Now let be $\phi = Z \circ \theta \in \mathrm{Hom}_{\mathbb{R}}(T^1_0(M), T^0_0(M))$ where $Z \in T^1_0(M)$ and $\theta \in T^0_1(M)$. For any $X \in T^1_0(M)$, $\phi(X) = Z(\theta(X)) \in T^0_0(M)$.

So the liftings of the vector fields and the differential forms determine the lifts of ϕ as follows.

$$\phi^V(X^C) = (\phi(X))^V = ((Z\circ\theta)(X))^V = (Z(\theta(X)))^V = Z^C(\theta(X))^V$$

$$= Z^C(\theta^V(X^C)) = (Z^C\circ\theta^V)(X^C) \quad \Longrightarrow \quad \phi^V = (Z\circ\theta)^V = Z^C\circ\theta^V .$$

As $\phi^V(X^V) = Z^C(\theta^V(X^V)) = Z^C(0) = 0$, the second condition is also verified.

Analogously,

$$\phi^V(X^C) = (\phi(X))^V = ((Z\circ\theta)(X))^V = (Z(\theta(X)))^V = Z^V(\theta(X))^C$$

$$= Z^V(\theta^C(X^C)) = (Z^V\circ\theta^C)(X^C) .$$

So there is a second vertical lift,

$$(\phi^V)' = Z^V\circ\theta^C ,$$

with $(\phi^V)'(X^V) = 0$.

With regard to the complete lift, $\phi^C(X^C) = (\phi(X))^C = (Z(\theta(X)))^C =$
$= Z^C(\theta(X))^C = Z^C(\theta^C(X^C)) = (Z^C \circ \theta^C)(X^C) \Longrightarrow \phi^C = (Z\circ\theta)^C = Z^C \circ \theta^C$,
and similarly we see that

$$(Z\circ\theta)^C(X^V) = (Z^C\circ\theta^C)(X^V) = Z^C(\theta^C(X^V)) = Z^C(\theta(X))^V =$$
$$= (Z(\theta(X)))^V = (Z\circ\theta)(X) .$$

So for a derivation $D:(dx^i;\phi^i_j)$ in M , with components $\phi^i_j = Z_j\circ\theta^i$, the complete lift is

$$D^C \quad : \quad \left(\begin{array}{c|cc} (dx^i)^V & (Z_j\circ\theta^i)^V & \bar{\phi}^i_{n+j} \\ (dx^i)^C & (Z_j\circ\theta^i)^C & (Z_j\circ\theta^i)^V \end{array}\right)$$

$$= \left(\begin{array}{c|cc} (dx^i)^V & (Z_j)^C\circ(\theta^i)^V & \bar{\phi}^i_{n+j} \\ (dx^i)^C & (Z_j)^C\circ(\theta^i)^C & (Z_j)^V\circ(\theta^i)^C \end{array}\right)$$

Notice that

$$((Z_j)^V\circ(\theta^i)^V)(X^C) = (Z_j)^V((\theta^i)^V(X^C)) = (Z_j)^V(\theta^i(X))^V = 0$$

so by symmetry we can define the complete lift of D in this particular case as

$$D^C \quad : \quad \left(\begin{array}{c|cc} (dx^i)^V & (Z_j)^C\circ(\theta^i)^V & (Z_j)^V\circ(\theta^i)^V \\ (dx^i)^C & (Z_j)^C\circ(\theta^i)^C & (Z_j)^V\circ(\theta^i)^C \end{array}\right)$$

This choice has a justification. Let us suppose that, in a local basis, is $Z_j = -\dfrac{\partial}{\partial x^j}$, $\theta^i = dx^i$. The derivation

$$D \quad : \quad (dx^i : (-\partial/\partial x^j)\circ dx^i)$$

is just the Lie derivation. In fact,

$$D_X \quad : \quad (X^i : (-\partial/\partial x^j)(dx^i(X))) = (X^i ; -\partial X^i/\partial x^j) ;$$

but, precisely,

$$L_X(\partial/\partial x^j) = [X,\ \partial/\partial x^j] = [(\partial/\partial x^i)\cdot X^i,\ \partial/\partial x^j]$$

$$= (\partial/\partial x^i)\cdot(-\ \partial X^i/\partial x^j) = (\partial/\partial x^i)\cdot Q^i_j$$

with $Q^i_j = \phi^i_j(X)$, hence :

$$L\ :\ (dx^i;\ (-\partial/\partial x^j)\cdot dx^i)\ .$$

Its complete lift to the tangent bundle will be, in this case :

$$L^C\ :\ \left(\begin{array}{c|cc} (dx^i)^V & (-\partial/\partial x^j)^C\circ(dx^i)^V & (-\partial/\partial x^j)^V\circ(dx^i)^V \\ (dx^i)^C & (-\partial/\partial x^j)^C\circ(dx^i)^C & (-\partial/\partial x^j)^V\circ(dx^i)^C \end{array}\right)$$

$$= \left(\begin{array}{c|cc} (dx^i)^V & (-\partial/\partial x^j)\circ dx^i & (-\partial/\partial x^{n+j})\circ\ dx^i \\ (dx^i)^C & (-\partial/\partial x^j)\circ dx^{n+i} & (-\partial/\partial x^{n+j})\circ dx^{n+i} \end{array}\right)$$

that is, the Lie derivation in TM . It is interesting to observe that, for the incorporation of this case, is inconvenient to impose "a priori" the excessively simplist condition $\bar{\phi}^i_{n+j} = 0$, as we proposed in [2].

6.- Let us consider $\phi = \theta\circ S$, with $\theta \in T^o_1(M)$ and $S \in T^1_1(M)$, considered as an endomorphism of $T^1_0(M)$. In a way analogous to the last case we obtain :

$$(\theta\circ S)^V = \theta^V\circ S^C = \theta^C\circ S^V\ ,\quad (\theta\circ S)^C = \theta^C\circ S^C\ .$$

If a derivation $D : (dx^i;\phi^i_j)$ has $\phi^i_j = \omega^i_j\circ S$, $\omega^i_j \in T^o_1(M)$, a similar method to one used before gives

$$D^C\ :\ \left(\begin{array}{c|cc} (dx^i)^V & (\omega^i_j)^C\circ S^V & (\omega^i_j)^V\circ S^V \\ (dx^i)^C & (\omega^i_j)^C\circ S^C & (\omega^i_j)^V\circ S^C \end{array}\right)$$

where

$$((\omega^i_j)^V\circ S^V)(X^C) = (\omega^i_j)^V(S^V(X^C)) = (\omega^i_j)^V(SX)^V = 0\ ,$$

verifying the required condition for $\bar{\phi}^i_{n+j} = (\omega^i_j)^V\circ S^V$.

Notwithstanding the most interesting lift proceeds from the following consideration: Given the covariant derivation $\nabla : (dx^i; \omega^i_j)$, already $\nabla \circ S : (dx^i \circ S; \omega^i_j \circ S)$ is not a derivation but a pseudoconnection with tensor S, as we have described in [1]. The lift obtained in the prior manner involves tensor fields S^V and S^C which impede to have a new pseudoconnection. Instead, from motivation like the precedent one, we impose the condition on the complete lift of being also a pseudoconnection: exactly the pseudoconnection with tensor S^C, the only existent possibility. First, it is verified

$$(dx^i \circ S)^V = (dx^i)^V \circ S^C , \quad (dx^i \circ S)^C = (dx^i)^C \circ S^C .$$

For the $\bar{\phi}$ we take

$$\bar{\phi}^i_j = (\omega^i_j)^V \circ S^C = \bar{\phi}^{n+i}_{n+j} , \quad \bar{\phi}^{n+i}_j = (\omega^i_j)^C \circ S^C ,$$

which satisfy the solicited conditions. With regard to $\bar{\phi}^i_{n+j}$, it will be of the type $\xi \circ S^C$, where ξ is a differential form; since $\bar{\phi}^i_{n+j}(X^C) = 0$, we have $\xi(S^C(X^C)) = \xi(SX)^C = 0$, therefore $\xi = 0$, as before we have seen, that is $\bar{\phi}^i_{n+j} = 0 \circ S^C = 0$. Thus, the complete lift of the pseudoconnection $\nabla \circ S$ is the pseudoconnection

$$(\nabla \circ S)^C \quad : \quad \left(\begin{array}{c|cc} (dx^i)^V \circ S^C & (\omega^i_j)^V \circ S^C & 0 \\ (dx^i)^C \circ S^C & (\omega^i_j)^C \circ S^C & (\omega^i_j)^V \circ S^C \end{array} \right)$$

that we obtained in [2] by the imposition $(\nabla \circ S)^C = \nabla^C \circ S^C$.

7.- Extensions of this case are possible. For instance, the pseudoderivations $D \circ S : (dx^i \circ S; \phi^i_j \circ S)$; or the case whose ϕ^i_j are equal to $\omega^i_k \circ S^k_j$, $\omega^i_k \in T^o_1(M)$, $S^k_j \in T^1_1(M)$. In the most important cases the method attaines a unique complete lift of a derivation (or an operator induced for it) by imposing that this lift and the derivation of which it proceeds are of the same type.

REFERENCES

[1] J.J. Etayo: "Pseudoderivaciones", Revista Matemática Hispanoamericana, vol. 35 (1975), 81-98.

[2] J.J. Etayo: "Lifts of derivations and differentiations to the tangent bundle", Proceedings of the IV International Colloquium of Differential Geometry, Santiago de Compostela (1979), 117-130.

[3] J.J. Etayo; "On a complete lifting of derivations", Tensor, vol. 38 (1982), 169-178.

[4] K. Yano and S. Ishihara: Tangent and cotangent bundles: Differentia Geometry, M. Dekker, New York (1973).

SOME EXAMPLES OF DEFORMATIONS OF TRANSVERSELY HOLOMORPHIC FOLIATIONS

J. Girbau

Secció de Matemàtiques de la
Universitat Autònoma de Barcelona.

Bellaterra (Barcelona) SPAIN

A transversely holomorphic foliation F on a smooth manifold X is given by an open covering $\{U_i\}_{i\in I}$ of X and differentiable submersions $f_i: U_i \longrightarrow \mathbb{C}^n$ such that for each $i,j \in I$ there is a holomorphic isomorphism g_{ij} of $f_j(U_i \cap U_j)$ on $f_i(U_i \cap U_j)$ such that $f_i = g_{ij} \circ f_j$. A typical example of this situation is the foliation on S^{2n+1} whose leaves are the fibres of the Hopf fibration $S^{2n+1} \longrightarrow P_n(C)$.

This note deals with the theory of deformations of such foliations.

The general theory of deformations of complex compact manifolds was initiated by Kodaira and Spencer [5] and completed by Kuranishi [7] who proved the existence of a germ of analytic space parametrizing a versal deformation for a given complex structure. Kodaira and Spencer themselves extended the theory to many other structures, in particular to complex analytic foliations on compact complex manifolds [6].

For smooth foliations, the theory does not apply at all, but for transversely holomorphic foliations on a compact manifold X, Duchamp and Kalka proved a weaker form of the Kuranishi theorem [2]. Haefliger, Sundararaman and the author proved a stronger version of this theorem [3] and gave specific examples and applications of it.

In this note I offer an alternative proof of this theorem relating the two points of wiew of [2] and [3]. Also I report on some examples of the joint paper with Haefliger and Sundararaman [3].

By technical reasons we have to deal with non reduced analytic spaces. We begin by recalling the fundamental definitions we need.

1. Non reduced analytic spaces (See [8]).

Suppose that U is an open set in $\mathbb{C}^n$. Let $\mathcal{O}_U$ be the sheaf of germs of complex analytic functions on U. Let I be a coherent sheaf of ideals of $\mathcal{O}_U$. Let V be the support of $\mathcal{O}_U/I$. Denote by $\mathcal{O}_V$ the restriction of $\mathcal{O}_U/I$ to V. The pair $(V, \mathcal{O}_V)$ is called a local model. A non reduced analytic space or, shortly, an analytic space is a pair $(X, \mathcal{O}_X)$, where X is a topological space and $\mathcal{O}_X$ a sheaf of local $\mathbb{C}$-algebras on X, which is locally isomorphic to a local model.

Simple examples.- 1) Denote by (x,y) the canonical coordinates of $\mathbb{C}^2$. If we take the ideal I of $\mathcal{O}_{\mathbb{C}^2}$ generated by x, the support V of $\mathcal{O}_{\mathbb{C}^2}/I$ is the line $x = 0$. In this case $\mathcal{O}_V \subset \mathcal{C}_V$ ($\mathcal{C}_V$ = the sheaf of germs of continuous functions on V).

2) If we take the ideal I of $\mathcal{O}_{\mathbb{C}^2}$ generated by x^2, the support V of $\mathcal{O}_{\mathbb{C}^2}/I$ is the line $x = 0$, too. But in this case $\mathcal{O}_V \not\subset \mathcal{C}_V$. These are examples of different non reduced analytic spaces on the line $x = 0$.

2. The fundamental sheaf associated to a foliation and its resolution.

Let F be a transversely holomorphic foliation on X given by a covering $\{U_i\}$ and submersions $f_i : U_i \longrightarrow \mathbb{C}^n$. Suppose that we have coordinates (x^u, x^a) on U_i such that the submersion f_i is expressed $(x,z) \longrightarrow z$. The fundamental sheaf Θ associated to F is the sheaf of germs of vector fields which in local coordinates are expressed

$$\xi = \xi^a(z) \frac{\partial}{\partial z^a},$$

where $\xi^a(z)$ are holomorphic. Also we shall need the sheaf Θ' of germs of smooth vector fields ξ on X which in the above local coordinates are of the form

$$\xi = \xi^u \frac{\partial}{\partial x^u} + \xi^a(z) \frac{\partial}{\partial z^a} + \xi^{\bar{a}} \frac{\partial}{\partial \bar{z}^a},$$

where the ξ^a depend only on the z's and the dependence is holomorphic.

Denote by F the subbundle which in each flat local chart (U, x^u, z^a) is generated by $\{\partial/\partial x^u, \partial/\partial \bar{z}^a\}$. F is independent of the choice of the flat local chart. Denote by I_F the ideal in the algebra ${}^cA^*(X)$ (of C^∞ complex-valued differential forms on X) consisting of those forms whose restriction to F vanish. I_F is locally generated by $\{dz^a\}$. Denote by A^*_F the quotient ${}^cA^*(x)/I_F$. The elements of A^*_F will be called F-differential forms. A^*_F is locally generated by the classes $<dx^u>$, $<d\bar{z}^a>$. Denote by D^q the space of complex derivations of degree q of A^*_F. Since I_F is closed under the exterior derivative d of ${}^cA^*(x)$, d gives a derivation d_F of A^*_F. An element $\delta \in D^q$ is represented in flat local coordinates (x^u, z^a) by the pair (φ,ξ) of vector-valued F-differential forms of degrees q and q + 1

$$\varphi = \varphi^u \frac{\partial}{\partial x^u} + \varphi^a \frac{\partial}{\partial z^a} \quad \varphi^{\bar{a}} \frac{\partial}{\partial \bar{z}^a}$$

$$\xi = \xi^u \frac{\partial}{\partial x^n} + \xi^{\bar{a}} \frac{\partial}{\partial \bar{z}^a}$$

where $\varphi^u = \delta x^u$, $\varphi^a = \delta z^a$, $\varphi^{\bar{a}} = \delta \overline{z^a}$, $\xi^u = (-1)^p <dx^u>$, $\xi^{\bar{a}} =$ $= (-1)^p\ \delta <d\ \bar{z}^a>$, where < > denotes "class in A^*_F/I_F".

If δ and δ' are derivations of A^*_F, one defines its Poisson's bracket by

$$[\delta,\ \delta'] = \delta\ \delta' - (-1)^{pq}\ \delta'\ \delta$$

where p and q are the degrees of δ and δ' respectively. If $u \in D^p$ one defines $D\,u = [d_F,\ u]$. If δ is expressed (φ,ξ) in a local chart U, then $D\,\delta$ is expressed $(d_F\ \varphi - \xi,\ -\ d\ \xi)$.

Denote by Ω'^q the sheaf of gems of elements of D^q. We have a sequence of sheaves

$$(1) \qquad 0 \longrightarrow \Theta' \longrightarrow \Omega'^0 \xrightarrow{D} \Omega'^1 \xrightarrow{D} \ldots\ldots$$

One can verify the following facts:

(i) This sequence is exact.

(ii) Ω'^q is a sheaf of modules over the ring of germs of differentiable functions on X and it is locally free (immediate in local coordinates). So Ω'^q is the sheaf of germs of smooth sections of a vector bundle E^q.

(iii) The symbol sequence

$$(2) \qquad 0 \longrightarrow p^*(E^0) \longrightarrow p^*(E^1) \longrightarrow \ldots\ldots$$

associated to (1) (where $p : {}^cT(X) - \{0\} \longrightarrow X$ is the natural projection) is exact except in $p^*(E^0)$. This can be easily proven in local coordinates.

Also we need a resolution of Θ which has been considered by Duchamp and Kalka. Denote by ν^c the complex transverse bundle. We have a splitting ${}^c\nu = \nu^{1,0} \oplus \nu^{0,1}$. Let $A^q_F(\nu)$ be the space of C^∞ $\nu^{1,0}$-valued F-forms. If $\varphi \in A^q_F$ then φ is represented by $\varphi = \varphi^a \langle \partial/\partial z^a \rangle$ in a coordinate system, where φ^a is an F-form and $\langle \partial/\partial z^a \rangle$ denotes the class of $\partial/\partial z^a$ in ν. We define an operator $d_F : A^q_F(\nu) \longrightarrow A^{q+1}_F(\nu)$ by $d_F\,\varphi = d_F\,\varphi^a \langle \partial/\partial z^a \rangle$. It is easy to see that d_F is a global operator.

Denote by Ω^q the sheaf of germs of elements of $A^q_F(\nu)$. The sequence

$$0 \longrightarrow \Theta \longrightarrow \Omega^0 \longrightarrow \Omega^1 \longrightarrow \ldots\ldots$$

is exact. The canonical projection $\pi : {}^cT(X) \longrightarrow \nu^{1,0}$ induces a projection $\pi : \Theta' \longrightarrow \Theta$. We can define a projection $\pi : D^q \longrightarrow A^q_F(\nu)$ by $\pi(\delta)(v) = \pi\,\varphi(\nu)$ for all $v \in F$, where φ is the global form defined by the local expression $\delta = (\varphi, \xi)$. (One can see that $\varphi_i = \varphi_j$ on $U_i \cap U_j$). From the local expression of the derivative D it follows that the following diagrame is commutative

$$\begin{array}{ccccccccc}
0 & \longrightarrow & \Theta' & \longrightarrow & \Omega'^0 & \xrightarrow{D} & \Omega'^1 & \xrightarrow{D} & \ldots\ldots \\
 & & \downarrow{\scriptstyle\pi} & & \downarrow{\scriptstyle\pi} & & \downarrow{\scriptstyle\pi} & & \\
0 & \longrightarrow & \Theta & \longrightarrow & \Omega^0 & \xrightarrow{d_F} & \Omega^1 & \xrightarrow{d_F} & \ldots\ldots
\end{array}$$

Since the $\ker\{\Theta' \xrightarrow{\pi} \Theta\}$ is a fine sheaf, π induces an isomorphism $H^p(X,\Theta') \longrightarrow H^p(X,\Theta')$ for all $p \geqslant 1$. ($H^o(X,\Theta)$ is finite dimensional, but $H^o(X,\Theta')$ is not).

3. Deformations and the generalized Kodaira-Spencer-Kuranishi theorem.

Let X be a differentiable manifold endowed with a transversely holomorphic foliation F. Let S be an analytic space with a distinguished point $0 \in S$. A deformation F^s of F parametrized by $(S,0)$ is given by a covering $\{U_i\}$ of X and smooth families of submersions $f^s_i : U_i \longrightarrow \mathbb{C}^n$, parametrized by $(S,0)$, holomorphic in s for fixed

$x \in U_i$, such that

$$f_i^s = g_{ij}^s \circ f_j^s \quad \text{on} \quad U_i \cap U_j,$$

where g_{ij}^s is a holomorphic family of $\mathbb{C}$-analytic isomorphisms of $f_j^s(U_i \cap U_j)$ on $f_i^s(U_i \cap U_j)$ (so holomorphic in both variables $(x,z) \in (S,o) \times \mathbb{C}^n$). Moreover the given foliation F is F^o, namely it can defined by the local submersions f_i^o.

In an analogous way we can define a germ of deformation F parametrized by a germ of analytic space (S,o).

The generalized Kodaira-Spencer-Kuranishi theorem can be stated as follows.

Theorem.- *Let F be a transversely holomorphic foliation on a compact manifold X. Denote by Θ its fundamental sheaf. There is a germ of deformation F^s of F parametrized by (S,o) such that for any germ of deformation $\tilde{F}^t$ of F parametrized by a germ of analytic space (T,o) there is an analytic map $\varphi:(T,o) \longrightarrow (S,o)$ so that the deformation $F^{\varphi(t)}$ induced by φ, parametrized by (T,o), is isomorphic to $\tilde{F}^t$. The tangent map $(d\varphi)_o$ of φ at o is unique. Moreover (S,o) is the germ at o of an analytic subspace of $H^1(X,\Theta)$ and $T_oS = H^1(X,\Theta)$. (S,o) is called the Kuranishi space of F and F^s the versal deformation of F.*

For the proof of the theorem we shall need the proposition stated in the following section.

4. A preliminary proposition.

Let F be a transversely holomorphic foliation on X. Denote by m the dimension of X. Let F^s be a real analytic deformation of F parametrized by an analytic space S with distinguished point o. Denote by N the disjoint union $\bigcup_{s \in S} \nu_s^{1,0}$ where $\nu_s^{1,0}$ is the complex transverse bundle of type $(1,0)$ associated to F^s. We have a natural projection $\pi: N \longrightarrow S$. Let Γ^S be the pseudogroup of real analytic automorphisms of $S \times \mathbb{R}^m \times \mathbb{C}^n$, $g(s,x,z)$, complex analytic in s and z, commuting with the projections on S and $\mathbb{R}^m$. In a natural way N can be endowed with a structure of real analytic space, together with a structure of Γ^S- manifold such that the local charts commute with the projections on S and $\mathbb{R}^m$, and such that for every local chart (W,φ) of N and for fixed $x \in X$ and $s \in S$, the restriction of φ to the fibre $(\nu_s^{1,0})_x$ of $\nu_s^{1,0}$ over x gives a $\mathbb{C}$-analytic isomorphism onto its image (in $\mathbb{C}^n$).

We have the following.

Propositon.- *Denote by γ_s the zero section of $\nu_s^{1,0}$. There is a neighbourhood S' of 0 in S, a neighbourhood U of the subset $\{\gamma_s(x), s \in S', x \in X\}$ of N, and a smooth map $g : U \longrightarrow X$ such that*

(i) $g(\gamma_s(x)) = x$ for $s \in S'$

(ii) For $s \in S'$ and $x \in X$ denote by $g_{s,x}$ the restriction of g to $(\nu_s^{1,0})_x \cap U$. The differential of $g_{s,x}$ at $\gamma_s(x)$ is the identity on $(\nu_s^{1,0})_x$.

(iii) For each $s \in S'$, $g_{s,x}$ is a $\mathbb{C}$-analytic isomorphism of $(\nu_s^{1,0})_x \cap U$ onto its image, with the complex structure induced by the transversely holomorphic foliation F^s.

Proof. Let F be the sheaf on N of germs of local C^∞ mappings of N into X verifying the conditions of the proposition. This sheaf is locally soft ("mou"), so also globally. Hence this sheaf has a section. To conclude, apply theorem 3.3.1, page 150, in Godement [4].

5. Proof of the theorem.

Without loss of generality we can assume that X is a real analytic manifold and that the transversely holomorphic foliation F is also real analytic (because any transversely analytic foliation is isotopic to a real analytic foliation).

We are going to relate the transversely holomorphic foliations $\widetilde{F}$ near F to the sections of Ω'^1 and Ω^1 (we use the notations of section 1). Let δ be an element of D^1, $\delta = (\varphi_i, c_i)$ on each local chart U. The first components φ_i give a global vector-valued F-form, because $\varphi_i = \varphi_j$ on $U_i \cap U_j$. Suppose that φ gives a monomorphism $F_x \longrightarrow {}^cT_x(X)$ at each point $x \in X$ and that the map

$$F_x \times F_x \longrightarrow {}^cT_x(X)$$

$$(v, w) \longrightarrow \varphi_x(v) + \overline{\varphi_x(w)}$$

is surjective. When $\delta = d_F$ these properties hold, so they will be fulfilled by the derivations δ near d_F. Set $\widetilde{F} = \varphi(F)$.

We have

$$\widetilde{F} + \overline{\widetilde{F}} = {}^{c}T(x) \ .$$

Take a flat local chart (U, x^u, z^a) and denote by x^α all the coordinates $x^u, \overline{z^a}$. Suppose that in this local chart $\delta = (\varphi, c)$ where c has an expression

$$c = \frac{1}{2} c^{\gamma}_{\alpha\beta} < dx^{\alpha} > \wedge < dx\ \beta > \otimes \frac{\partial}{\partial\ x^{\gamma}} \ .$$

As in [6] page 71, one can verify that the condition $[\delta,\delta] = 0$ is equivalent to

$$[\varphi_\alpha, \varphi_\beta] = c^{\gamma}_{\alpha\beta} \ \varphi_\gamma$$

where $\varphi_\alpha = \varphi(\partial/\partial\ x^\alpha)$ and $[\varphi_\alpha, \varphi_\beta]$ means the bracket of vector fields. As the φ_α are a basis of $\widetilde{F}$ we see that $[\delta,\delta] = 0$ is equivalent to $[\widetilde{F}, \widetilde{F}] \subset \widetilde{F}$. If such condition is fulfilled we have a transversely holomorphic foliation $\widetilde{\mathcal{F}}$ given by $\widetilde{F}$, by the complex Frobenius theorem. If we set $\delta = d_F + \omega$, the integrability condition $[\delta,\delta] = 0$ can be written in terms of ω by

$$D\ \omega - \frac{1}{2} [\omega, \omega] = 0 \ .$$

Now fix a real analytic riemannian metric on X. Denote by L the subbundle of $T(x)$ consisting of vectors tangent to the leaves of $\mathcal{F}$. We have splittings $T(x) = L \oplus \nu$, ${}^{c}T(x) = {}^{c}L \oplus \nu^{1,0} \oplus \nu^{0,1}$, $F = {}^{c}L \oplus \nu^{0,1}$.

Denote by π_F and $\pi_{1,0}$ the projections of ${}^{c}T(x)$ onto F and $\nu^{1,0}$ given respectively by the splitting above. If $\widetilde{\mathcal{F}}$ is a transversely holomorphic foliation whose space F is denoted by $\widetilde{F}$, we shall say that $\widetilde{\mathcal{F}}$ is near $\mathcal{F}$ if π_F maps $\widetilde{F}$ onto F isomorphically. If such is the case, let φ be the map

$$F \xrightarrow{\pi_F^{-1}} \widetilde{F} \xrightarrow{\pi_{1,0}} \nu^{1,0}$$

$\varphi \in A^1_F(\nu)$. $\widetilde{F}$ can be described by means of φ as $F = \{ v + \varphi(v), v \in F \}$, and we have $[\widetilde{F}.\widetilde{F}] \subset \widetilde{F}$. This integrability condition can be written in terms of φ as Ducham and Kalka do [2]. To state this condition we need some definitions. If $\varphi, \psi \in A^1_F(\nu)$ we denote by $[\varphi, \psi]$

the element of $A_F^2(\nu)$ defined by

$$2[\varphi,\psi](v_1,v_2) = \varphi\,\pi_F\,[v_1,\psi(v_2)] - \varphi\,\pi_F\,[v_2,\psi(v_1)]$$

$$+ \psi\,\pi_F\,[v_1,\varphi(v_2)] - \psi\,\pi_F\,[v_2,\varphi(v_1)]$$

$$- \pi_{1,0}\,[\varphi(v_1),\psi(v_2)] - \pi_{1,0}\,[\varphi(v_2),\psi(v_1)].$$

Denote by $P(\varphi)$ the element of $A_F^2(\nu)$ defined by

$$P(\varphi)(v_1,v_2) = \varphi\,\pi_F\,[\varphi(v_1),\varphi(v_2)]$$

With these notations the condition $[\widetilde{F},\widetilde{F}] \subset \widetilde{F}$ is equivalent to

$$(3) \qquad d_F\,\varphi + [\varphi,\varphi] + P(\varphi) = 0$$

We have now all the ingredients to adapt the Donady's proof of Kuranishi's theorem [1] to our situation.

Denote by ${}^r\mathcal{D}^q$ the space of C^r-sections of the fibre bundle E^q which appears in the sequence (2). ${}^r\mathcal{D}^q$ is a Banach space. We shall take a non integer $r>1$. Take real analytic metrics in each bundle E^q and denote by D^* the adjoint operator of D with respect to these metrics. Denote by θ the analytic map

$$ {}^r\mathcal{D}^1 \longrightarrow {}^{r-1}\mathcal{D}^2$$

$$\omega \longrightarrow D\omega - \frac{1}{2}[\omega,\omega].$$

The tangent map of θ at the origin is D. Since r is non integer and the sequence (2) is exact (except at $p^*(E^o)$) we have

$${}^r\mathcal{D}^1 = \operatorname{Im} D \oplus \ker D^*, \quad {}^{r-1}\mathcal{D}^2 = \operatorname{Im} D \oplus \ker D^*$$

Set $\Sigma = \{\omega \in {}^r\mathcal{D}^1$ such that $D^*\theta(\omega) = 0\}$. It is a $\mathbb{C}$-analytic Banach submanifold of ${}^r\mathcal{D}^1$ in a neighbourhood of the origin. Set $H = \Sigma \cap \ker D^*$. It is easy to see that $H = \{\omega \in {}^r\mathcal{D}^1$ such that $\omega = H\omega + \frac{1}{2} D^* G[\omega,\omega]\}$, where $H\omega$ denotes the harmonic part of ω with respect to the Laplace operator $\Delta = D^*D + D\,D^*$ and G denotes the Green's operator. As in [1], the elements of H are C^∞ and the tangent space of H at the origin is the space of harmonic derivations of degree 1, isomorphic to $H^1(X,\Theta')$. Set $S = \{\omega \in H$ such that $\theta(\omega) = 0\}$. To each $s \in S$ we associate the foliation F_s associated to the derivation $d_F + s$. So we have a family of foliations parametrized by S and we want to see that the germ of this family at the origin is versal. Denote by φ_s the element of $A^1_F(\nu)$ associated to F_s. φ_s fulfils condition (3). Denote by ${}^rA^1_F(\nu)$ the space of $\nu^{1,0}$ - valued F-forms of degree 1 and class C^r. Set $\phi^r = \{\varphi \in {}^rA^1_F(\nu)$ fulfilling (3)$\}$. Denote by ${}^{r+1}\Gamma(\nu)$ the space of C^{r+1} real sections of the real normal bundle ν. Denote by ρ the map

$$S \times {}^{r+1}\Gamma(\nu) \xrightarrow{\rho} \phi^r \quad {}^rA^1_F(\nu)$$

$$(s, \xi) \longrightarrow \varphi_s \circ g_\xi$$

where g_ξ means the diffeomorphism of X $x \longrightarrow g(\xi(x))$, where g is the map given by the proposition of section 4. ($\varphi_s \circ g_\xi$ means the element of ${}^rA^1_F(\nu)$ associated to the foliation whose space F is $(g_\xi)_*(F_s)$). The map ρ is only defined in a neighbourhood of the origin.

The orthogonal projection $T(X) \longrightarrow \nu$ gives (for small s) an isomorphism $\nu \cong \nu_s$ which gives a splitting ${}^c\nu \cong {}^c\nu_s \cong \nu^{1,0}_s \oplus \nu^{0,1}_s$. This induces a splitting of ${}^c({}^{r+1}\Gamma(\nu))$ which induces a family of complex structures on the Banach space ${}^{r+1}\Gamma(\nu)$ parametrized by a neighbourhood of the origin of S. We rename this neighbourhood by S. This family can be interpreted as a complex structure on $S \times {}^{r+1}\Gamma(\nu)$. The map ρ is $\mathbb{C}$-analytic in a neighbourhood of $(0,0)$.

The tangent map of $\rho_0 : \{0\} \times {}^{r+1}\Gamma(\nu) \longrightarrow {}^rA^1_F(\nu)$ at the origin is nothing but d_F.

Take a supplementary G of ker d_F in ${}^{r+1}\Gamma(\nu^{1,0})$ (ker d_F is finite -dimensional). Identify ν with $\nu^{1,0}$ and denote by $E = S \times G$. One can see that $(d\rho)_{(0,0)}$ maps isomorphically $T_{(0,0)}(E)$ onto $T_0(\phi^r) =$ $= \ker d^1_F \subset {}^rA^1_F(\nu)$.

Now we are able to adapt here the rest of the proof of the Kuranishi's theorem given in [1].

6. How to compute the Kuranishi space S.

In some concrete examples Haefliger, Sundararaman and the author have computed S, [3], by one of the two following ways:

1) By its equations, as in the following example: Let $\pi : X \longrightarrow T$ be a principal circle bundle over a complex torus $T = C/L$ with non zero Euler class. Let F be the transversely holomorphic foliation on X whose leaves are the fibres of π. The Kuranishi space of F is a non reduced space consisting of a double line in the 2-dimensional space $H^1(X, \Theta)$. The deformations of this foliation are the same that the deformations of the complex structure of the torus.

2) By constructing a versal deformation directly. In this way we have proven that the Kuranishi space of the foliation on S^{2n+1} whose leaves are the fibres of the Hopf fibration $S^{2n+1} \longrightarrow P_n(\mathbb{C})$ is a neighbourhood of the origin in the $((n + 1)^2 - 1)$-dimensional space $H^1(S^{2n+1}, \Theta)$.

REFERENCES

|1| A. Douady, "Le problème des modules pour les variétés analytiques complexes (d'après Kuranishi)". Seminaire Bourbaki n°227 (1964).

|2| T. Duchamp and M. Kalka, "Holomorphic foliations and deformations of the Hopf foliation", preprint.

|3| J. Girbau, A. Haefliger, D. Sundararaman, "On deformations of transversely holomorphic foliations", preprint IHES (1982). To be published in J. reine angew. Math.

|4| R. Godement, Théorie des faisceaux, Hermann.

|5| K. Kodaira and D.C. Spencer, "On deformations of complex analytic structures I, II, Ann. of Math. 67 (1958), 328-466; III, 71 (1960) 43-76.

|6| K. Kodaira and D.C. Spencer, "Multifoliate structures", Ann. of Math. 74 (1961), 52-100.

|7| M. Kuranishi, "Deformatons of compact complex manifolds", les Presses de l'Université de Montreal, 1971.

|8| B. Malgrange, "Analytic spaces", l'Enseignement Mathématique, 14 (1968), 1-28.

SUR CERTAINES EXPRESSIONS GLOBALES D'UNE FORME DE CONTACT

J. Gonzalo et F. Varela
Departamento de Matemáticas
Facultad de Ciencias
Universidad de Murcia.
Murcia. Spain.

Dans l'étude globale des formes de contact sur une variété compacte, on prend parfois le point de vue de mettre en face de la condition de contact, une condition accessoire sur la forme. Les exemples suivants mettent en évidence une telle situation.

1. Boothby-Wang, [2], considèrent des formes de contact vérifiant la propriété que les orbites de son système dynamique sont des cercles.

2. R. Lutz, [6], considère des formes de contact invariantes pour l'action d'un certain groupe de Lie.

3. Lutz-Hangan, [5], considèrent des formes de contact totalement géodésiques par rapport à une métrique de Riemann fixée sur la variété.

Une forme de contact sur une variété compacte de dimension 2n+1, est une forme de Pfaff vérifiant partout:

$$\omega \wedge (d\omega)^n \neq 0.$$

Toute forme de contact C^∞ (resp. C^ω) sur une variété compacte C^∞ (resp. C^ω) s'exprime globalement sous la forme

$$\omega = \sum_{i=1}^{N} f_i (g_i \, dg_{i+1} - g_{i+1} \, dg_i),$$

où f_i, g_i sont des fonctions différentiables globales de la variété en $\mathbb{R}$.

Exemples.-

a) Sur S^3 on considère la forme de contact $i^*(\omega)$, où

$$\omega = x_1 dx_2 - x_2 dx_1 + x_3 dx_4 - x_4 dx_3.$$

Ainsi $i^*(\omega)$ s'exprime de façon globale:

$$i^*(\omega) = f_1 df_2 - f_2 df_1 + f_3 df_4 - f_4 df_3.$$

b) Sur $S^2 \times S^1$ on considère la forme de contact $i^*(\omega)$, où

$$\omega = y_1 d\theta + y_2 dy_3 - y_3 dy_2,$$

avec $\sum_{i=1}^{3} y_i^2 = 1$. Ainsi, $i^*(\omega)$ s'exprime globalement:

$$i^*(\omega) = f(f_1 df_2 - f_2 df_1) + f_3 df_4 - f_4 df_3.$$

c) La forme de contact sur le tore T^3;

$$\omega = \cos\theta_1 d\theta_2 + \operatorname{sen}\theta_1 d\theta_3,$$

s'exprime à l'aide de fonctions globales sous la forme:

$$\omega = f(f_1 df_2 - f_2 df_1) + g(f_3 df_4 - f_4 df_3)$$

Dans [4] on montre le

Théorème 1: Soit M_3 une variété compacte, connexe de dimension trois, sur laquelle il existe une forme de contact, qui s'exprime globalement sous la forme $\omega = f(f_1 df_2 - f_2 df_1) + g(f_3 df_4 - f_4 df_3)$. Alors:

a) L'ensemble $f = 0$ (resp. $g = 0$) est vide ou bien réunion de tores T^2.

b) L'ensemble $f_1 = f_2 = 0$ (resp. $f_3 = f_4 = 0$) est vide ou bien réunion de cercles.

c) Si $\{f_1 = f_2 = 0\} \neq \emptyset$ et $f_3^2 + f_4^2 \neq 0$ partout, alors M_3 est difféomorphe à $S^1 \times S^2$ et $\{f_1 = f_2 = 0\}$ sont deux cercles enlacés.

d) Si $\{f_1 = f_2 = 0\} \neq \emptyset$ et $\{f_3 = f_4 = 0\} \neq \emptyset$, alors M_3 est difféomorphe à la sphère S^3 et $\{f_1 = f_2 = 0\}$, $\{f_3 = f_4 = 0\}$ sont deux cercles enlacés une fois.

e) Si $f_1^2 + f_2^2 \neq 0$ et $f_3^2 + f_4^2 \neq 0$ partout, alors M_3 est difféomorphe au tore T^3.

Remarque 1: Le théorème 1 exprime une liaison trés étroite entre l'expression globale d'une forme de contact et la structure différentiable de la variété.

Remarque 2: Dans les exemples a), b), c) l'ensemble $f = 0$ est vide pour

S^3, contient un tore pour $S^2 \times S^1$ et deux tores pour T^3.

Remarque 3: Le théorème 1 permet de construire sur S^3 des formes de contact ayant un nombre arbitraire de tores dans l'ensemble $f = 0$ à condition que le nombre de tores de $f = 0$ soit égal au nombre de tores de $g = 0$. Les formes de contact ainsi obtenues sur S^3 ne sont pas équivalentes comme structures de contact à celle définie sur S^3 par $\omega = x_1dx_2 - x_2dx_1 + x_3dx_4 - x_4dx_3$. Pour le démontrer, on constate que ω_n et ω_m, si n est pair et m impair, ne sont pas homotopes dans l'espace des formes de Pfaff sans zéros sur S^3. En particulier, d'après J. Cerf, [3], ω_n et ω_m ne sont pas isomorphes. D'ailleurs si $n \neq 0$, on constate l'existence de courbes intégrales fermées de l'équation de Pfaff $\omega_n = 0$, bordant un disque transverse à la forme. Le théorème de Bennequin, [1], permet de conclure la non isomorphie des structures ω_n et ω. Dans cette remarque n est le nombre de tores de $f = 0$.

Dans cet exposé, nous considérons des formes de contact sur une variété de dimension cinq, ayant la propriété d'avoir une expression globale fixée.

Exemples dans la dimension cinq.

a) Sur S^5 on considère la forme de contact $i^*(\omega)$, où $\omega = x_1dx_2 - x_2dx_1 + x_3dx_4 - x_4dx_3 + x_5dx_6 - x_6dx_5$. Ainsi $i^*(\omega)$ s'exprime sous la forme $i^*(\omega) = f_1df_2 - f_2df_1 + f_3df_4 - f_4df_3 + f_5df_6 - f_6df_5$.

b) Sur $S^1 \times S^4$ on considère la forme de contact $i^*(\omega)$, où $\omega = y_5d\theta + y_1dy_2 - y_2dy_1 + y_3dy_4 - y_4dy_3$ avec $\sum y_i^2 = 1$. Ainsi
$i^*(\omega) = f(f_1df_2 - f_2df_1) + f_3df_4 - f_4df_3 + f_5df_6 - f_6df_5$.

c) Sur $S^1 \times S^1 \times S^3$ on considère la forme de contact $i^*(\omega)$, où $\omega = y_1d\theta_1 + y_2d\theta_2 + y_3dy_4 - y_4dy_3$. Elle s'exprime
$i^*(\omega) = f(f_1df_2 - f_2df_1) + g(f_3df_4 - f_4df_3) + f_5df_6 - f_6df_5$.

d) Sur $S^1 \times S^1 \times S^1 \times S^2$ on prend la forme de contact $i^*(\omega)$, où $\omega = y_1d\theta_1 + y_2d\theta_2 + y_3d\theta_3$ avec $\sum y_i^2 = 1$. Elle admet une expression globale du type
$f(f_1df_2 - f_2df_1) + g(f_3df_4 - f_4df_3) + h(f_5df_6 - f_6df_5)$.

<u>Théorème 2</u>: Soit M_5 une variété compacte, connexe, de dimension cinq, avec une forme de contact qui s'exprime globalement sous la forme $\omega = f(f_1 df_2 - f_2 df_1) + f_3 df_4 - f_4 df_3 + f_5 df_6 - f_6 df_5$. Alors

a) Si $f_1^2 + f_2^2 \neq 0$ partout, M_5 est difféomorphe à $S^4 \times S^1$.

b) Si $\{f_1 = f_2 = 0\} \neq \emptyset$, cet ensemble est une S^3, la fonction f est différente de zéro partout, et M_5 est difféomorphe à la sphère S^5.

<u>Remarque 4</u>: Le théorème 2 montre en particulier que sur S^5 il n'y a pas de formes de contact du type $f(f_1 df_2 - f_2 df_1) + f_3 df_4 - f_4 df_3 + f_5 df_6 - f_6 df_5$ avec $f = 0$ sur un ensemble de S^5.

<u>Démonstration de la partie b) du théorème.</u>

Soient dans $\mathbb{R}^4 - \{0\}$ la trois forme

$$\Omega_0 = (x_1 dx_2 - x_2 dx_1) \wedge dx_3 \wedge dx_4 + (x_3 dx_4 - x_4 dx_3) \wedge dx_1 \wedge dx_2,$$

et le champ de vecteurs

$$X = x_1 \frac{\partial}{\partial x_1} + x_2 \frac{\partial}{\partial x_2} + x_3 \frac{\partial}{\partial x_3} + x_4 \frac{\partial}{\partial x_4} \quad .$$

Il est évident que Ω_0 est de rang constant égal à trois et son espace associé est engendré en chaque point p de $\mathbb{R}^4 - \{0\}$ par le vecteur X_p.

Soit $f: M_3 \longrightarrow \mathbb{R}^4 - \{0\}$ une application différentiable d'une variété compacte orientable de dimension trois sur $\mathbb{R}^4 - \{0\}$.

<u>Lemme 1</u>: Les deux propositions suivantes sont équivalentes:

a) $f^*(\Omega_0)$ est une forme de volume sur M_3;

b) f est une immersion transverse au champ X.

La preuve de ce lemme est très simple et nous en omettrons la démonstration.

<u>Lemme 2</u>: Soit M_3 une variété compacte, connexe, orientable de dimension trois avec une forme de volume qui s'exprime globalement sous la forme $\Omega = (f_1 df_2 - f_2 df_1) \wedge df_3 \wedge df_4 + (f_3 df_4 - f_4 df_3) \wedge df_1 \wedge df_2$. Alors, M_3 est difféomorphe à la sphère S^3.

<u>Preuve</u>: Soit $f: M_3 \longrightarrow \mathbb{R}^4 - \{0\}$ définie par $f(p) = (f_1(p), f_2(p), f_3(p), f_4(p))$. $f^*(\Omega_0)$ étant une forme de volume sur M_3, d'après le lemme 1, l'application f est transverse au champ X. Soit $\pi: \mathbb{R}^4 - \{0\} \longrightarrow S^3$

définie par $\pi(p) = p/|p|$. Le champ X étant transverse à la sphère $S^3 \subset \mathbb{R}^4$, l'application $\pi\circ f\colon M_3 \longrightarrow S^3$ est un difféomorphisme local de M_3 sur S^3 et alors un revêtement de S^3, d'où $\pi\circ f$ est un difféomorphisme, q.e.d.

Remarque 5: La condition de contact de la forme ω s'exprime

$$\omega \wedge (d\omega)^2 = 4f^2(f_1df_2 - f_2df_1) \wedge df_3 \wedge df_4 \wedge df_5 \wedge df_6 + \\ + 2df \wedge (f_1df_2 - f_2df_1) \wedge \left[(f_5df_6 - f_6df_5) \wedge df_3 \wedge df_4 + \right. \\ \left. + (f_3df_4 - f_4df_3) \wedge df_5 \wedge df_6\right] + 4fdf_1 \wedge df_2 \wedge \left[(f_3df_4 - f_4df_3) \wedge \right. \\ \left. \wedge df_5 \wedge df_6 + (f_5df_6 - f_6df_5) \wedge df_3 \wedge df_4\right] \neq 0$$

partout. Sur l'ensemble $f_1 = f_2 = 0$, on a:

$$4fdf_1 \wedge df_2 \wedge \left[(f_3df_4 - f_4df_3) \wedge df_5 \wedge df_6 + (f_5df_6 - f_6df_5) \wedge df_3 \wedge df_4\right] \neq 0,$$

d'où résulte le :

Lemme 3: L'ensemble $f_1 = f_2 = 0$ est une sous-variété régulière, compacte, de M_5 éventuellement non connexe, chaque composante connexe étant difféomorphe à S^3.

Remarque 6: Soit $Q = \{p \in M_5 \mid f(p) \neq 0\}$. La condition de contact implique $Q \neq \emptyset$. Sur Q, la structure de contact $\rho^2\omega$ peut s'exprimer sous la forme $\rho^2\omega = (\operatorname{sg} f)(h_1dh_2 - h_2dh_1) + (h_3dh_4 - h_4dh_3) + (h_5dh_6 - h_6dh_5)$, où

$$h_1 = |f|^{1/2}\rho f_1 \qquad h_3 = \rho f_3 \qquad h_5 = \rho f_5$$

$$h_2 = |f|^{1/2}\rho f_2 \qquad h_4 = \rho f_4 \qquad h_6 = \rho f_6$$

$$\rho = \left[|f|(f_1^2 + f_2^2) + f_3^2 + f_4^2 + f_5^2 + f_6^2\right]^{-1/2}.$$

On vérifie $\sum_{i=1}^{6} h_i^2 = 1$. Ainsi les h_i sont des fonctions de M_5 à $\mathbb{R}$, différentiables sur Q et continues sur M_5.

Remarque 7: Soit $A = \{p \in S^5 \mid x_1^2 + x_2^2 \neq 0\}$. L'application $\phi\colon A \longrightarrow S^1\times B_4$, où B_4 est la boule ouverte $\sum_{i=1}^{4} x_i^2 < 1$, donnée par

$$(x_1,x_2,x_3,x_4,x_5,x_6) \longrightarrow ((x_1^2+x_2^2)^{-1/2}x_1, (x_1^2+x_2^2)^{-1/2}x_2, x_3, x_4, x_5, x_6).$$

est un difféomorphisme. Ainsi $S^5 = S^3 \cup A$ avec $S^3 \cap A = \emptyset$. Soit H_i une composante connexe de l'ouvert $H = \{p \mid f(p) \neq 0,\ f_1^2(p)+f_2^2(p) \neq 0\}$, et soit $h: M_5 \longrightarrow S^5$ l'application continue définie par $h(p) = (h_i(p))$, $i : 1,\ldots,6$. Il est clair que $h(H) \subset A$.

Lemme 4:

a) $h|H_i : H_i \longrightarrow A$ est différentiable;

b) $h|H_i : H_i \longrightarrow A$ est un difféomorphisme local;

c) $h|H_i$ est un revêtement à un nombre fini de feuilles.

Preuve: a) est évident ; b) est une conséquence du fait que sur Q $\rho^2\omega = (h|Q)^*(\omega_0)$, où ω_0 est la forme de contact canonique sur la sphère S^5; c) est un conséquence de l'affirmation suivante: " soit p_n une suite de points de H_i tels que $h(p_n)$ soit convergente à un point q en A; alors p_n converge à un point $p \in H_i$ et $h(p) = q$". En effet, soit $q = \lim_{n\to\infty} h(p_n)$ Etant donné que $q \in A$, on a $\lim_{n\to\infty} (h_1(p_n), h_2(p_n)) \neq (0,0)$, et alors

$$\lim_{n\to\infty} (|f(p_n)|^{1/2}\rho(p_n)f_1(p_n), |f(p_n)|^{1/2}\rho(p_n)f_2(p_n)) \neq (0,0) \qquad (*)$$

M_5 étant compacte, soit p un point d'accumulation de la suite (p_n). De (*) et de la continuité de la fonction h on a:

$$\left.\begin{array}{lll} f(p) \neq 0 & \Longrightarrow & p \in Q \\ (f_1^2+f_2^2)(p) \neq 0 & \Longrightarrow & p \in H \end{array}\right\} \Longrightarrow p \in H_i \text{ et } f(p) = q, \text{ q.e.d.}$$

Soit A_1 une composante connexe de la sous-variété $f_1 = f_2 = 0$. La condition de contact implique $A_1 \subset Q$; d'après la remarque 5 et le lemme 2, A_1 est une S^3, $(h|A_1)(A_1) \subset S^3$, $h|A_1 : A_1 \longrightarrow S^3$ est un revêtement et alors h est un difféomorphisme.

Lemme 5: Pour chaque composante connexe A_i de $f_1 = f_2 = 0$, il existe une composante connexe unique H_i de H telle que $A_i \cup H_i$ est ouvert.

C'est une conséquence directe de la compacité de A_i et du fait que A_i est une sous-variété régulière.

D'après les lemmes antérieurs on a:

a) $h: A_1 \longrightarrow S^3$ est un difféomorphisme;

b) $h|H_1 : H_1 \longrightarrow A$ est un revêtement à un nombre fini de feuilles;

c) $h: A_1 \cup H_1 \longrightarrow S^3 \cup A = S^5$ est un difféomorphisme local;

d) $A_1 \cup H_1$ est ouvert.

<u>Lemme 6</u>: $h: H_1 \longrightarrow A$ est un difféomorphisme.

<u>Preuve</u>: Soit $p \in A_1$; alors $h(p)=(0,0,x_3,x_4,x_5,x_6) \in S^3$; comme h est un difféomorphisme local, il existe un voisinage $V(p) \subset A_1 \cup H_1$ et un voisinage $V'_{h(p)} \subset S^3 \cup A$ tel que $h|V(p): V(p) \longrightarrow V'_{h(p)}$ est un difféomorphisme. D'après la remarque 7, dans tout voisinage $V'_{h(p)}$ il existe un générateur γ de $\Pi_1(A) = \mathbb{Z}$. La courbe $(h|V(p))^{-1}(\gamma)$ vérifie:

a) $(h|V(p))^{-1}(\gamma) \subset V(p)$;

b) $(h|V(p))^{-1}(\gamma) \subset H_1$

c) $h((h|V(p))^{-1}(\gamma)) = \gamma$.

De a), b), c) on déduit que $h|H_1: H_1 \longrightarrow A$ est un homéomorphisme et alors un difféomorphisme, q.e.d.

Le théorème 2, b) résulte maintenant du fait que $A_1 \cup H_1$ étant ouvert et fermé, $M_5 = A_1 \cup H_1 = S^5$, q.e.d.

De façon analogue on démontre les théorèmes suivants :

<u>Théorème 3</u>: Soit M_5 une variété compacte, connexe, de dimension 5, avec une forme de contact que s'exprime globalement sous la forme

$$\omega = f(f_1 df_2 - f_2 df_1) + g(f_3 df_4 - f_4 df_3) + h(f_5 df_6 - f_6 df_5).$$

Alors M_5 est difféomorphe à une des variétés suivantes: S^5, $S^1 \times S^4$, $T^2 \times S^3$, $T^3 \times S^2$.

<u>Théorème 4</u>: Soit M_{2n+1} une variété compacte, connexe, avec une forme de contact qui s'exprime globalement sous la forme

$$\omega = f(f_1 df_2 - f_2 df_1) + \cdots + (f_{2n+1} df_{2n+2} - f_{2n+2} df_{2n+1}).$$

Alors, $f_1^2 + f_2^2 \neq 0$ partout implique $M_{2n+1} \approx S^1 \times S^{2n}$; et $\{f_1 = f_2 = 0\} \neq \emptyset$ implique $M_{2n+1} \approx S^{2n+1}$ et $f \neq 0$ partout.

<u>Démonstration de la partie a) du théorème 2</u>

On peut supposer $f_1^2 + f_2^2 = 1$ et $f^2 + f_3^2 + f_4^2 + f_5^2 + f_6^2 = 1$. Soit $h: M_5 \longrightarrow S^1 \times S^4$ l'application différentiable définie par

$$h(p) = (f_1(p), f_2(p), f_3(p), f_4(p), f_5(p), f_6(p)),$$

et soit ω_0 la forme de contact sur $S^1 \times S^4$ de l'exemple b). On a $h^*(\omega_0) = \omega$ et alors $h^*(\omega_0 \wedge (d\omega_0)^2) = \omega \wedge (d\omega)^2$, d'où h est de rang constant égal à 5. En conséquence (h, M_5) est un revêtement compact de $S^1 \times S^4$, d'où

M_5 est difféomorphe à $S^1 \times S^4$, q.e.d.

REFERENCES

[1] D. Bennequin, Entrelacements et équations de Pfaff, thèse.

[2] W.M. Boothby and H.C.Wang, On contact manifolds, Ann. of Math. 68 (1958), 731-734.

[3] J. Cerf, Sur les difféomorphismes de S^3, Lect. Notes in Mathematics 53, Springer, 1968.

[4] J. Gonzalo et F. Varela, Modèles globaux d'une variété de contact, à apparaître dans Astérisque.

[5] T. Hangan et R. Lutz, Formes de contact totalement géodésiques, Conférences de T. Hangan données à l'Université de Murcia.

[6] R. Lutz, Structures de contact sur les fibrés principaux en cercles de dimension 3, Ann. Inst. Fourier, 27 (1977).

CONNEXIONS SINGULIERES ET CLASSE DE MASLOV

J. Grifone et E. Hassan
Groupe d'Analyse Globale
et Physique Mathématique.
U.E.R. de Mathématiques.
118 route de Narbonne.
31062 Toulouse - Cedex.
(France)

INTRODUCTION-

On considère les solutions des équations différentielles du type $A(x,\dot{x})\ddot{x} + \Gamma(x,\dot{x}) = 0$ où A est une matrice pouvant dégénérer, comme des "géodésiques" de "connexions singulières". Par des méthodes de géométrie symplectique on associe à certaines connexions singulières une classe de cohomologie entière qui est nulle si la connexion est régulière (connexion au sens classique). La classe apparaît ainsi liée à la dégénérescence de la connexion. Des applications sont données qui mettent en évidence que la classe est liée à une théorie du dégré des "géodésiques" et permet, entre autres choses, une interprétation symplectique de la courbure des courbes planes et de l'indice d'un champ de vecteurs dans le plan.

1.- NOTATIONS ET DEFINITIONS.

Les notations sont celles de [3]. En particulier J désigne l'endomorphisme vertical, c'est-à-dire la structure presque-tangente naturelle de TM. Les connexions y sont caractérisées par la structure presque produit Γ sur TM vérifiant $J\Gamma = J$ et $\Gamma J = - J$.

Remarquons qu'une connexion peut être aussi définie par le projecteur vertical $1/2\ (I - \Gamma)$ caractérisé par les relations

$$Jv = 0 \quad \text{et} \quad vJ = J$$

Comme dans [3] on note F la structure presque-complexe sur TM associée à Γ, caractérisée par $Fh = - J$ et $FJ = h$ où $h = 1/2\ (I + \Gamma)$ est le projecteur horizontal.

Si VTM désigne le fibré vertical, on notera, pour $z \in TM$

$$i_z : V_z TM \longrightarrow T_{\pi(z)}M$$

l'isomorphisme naturel, ainsi que l'isomorphisme entre les algèbres tensorielles correspondantes.

Pour abréger les notations, si $\dot{t}$ est un tenseur sémi-basique sur TM on posera

$$t_{(z)} = i_z \dot{t}$$

En coordonnées locales adaptées à la fibration $\pi : TM \longrightarrow M$, si par exemple

$$\dot{t} = t^{\lambda}_{\alpha_1 \ldots \alpha_p}(x,y)\, dx^{\alpha_1} \wedge \ldots \wedge dx^{\alpha_p} \otimes \frac{\partial}{\partial y^{\lambda}}$$

on a,

$$t_{(z)} = t^{\lambda}_{\alpha_1 \ldots \alpha_p}(x,z)\, dx^{\alpha_1} \wedge \ldots \wedge dx^{\alpha_p} \otimes \frac{\partial}{\partial x^{\lambda}}$$

Si $\dot{A}$ et $\dot{B}$ sont deux tenseurs $\binom{1}{1}$ sémi-basiques sur TM, $\dot{A}F\dot{B}$ ne dépend pas du choix de la connexion et sera noté $\dot{A} \circ \dot{B}$. En coordonnées locales, si $\dot{A} = A^{\lambda}_{\alpha}\, dx^{\alpha} \otimes \frac{\partial}{\partial y^{\lambda}}$ et $\dot{B} = B^{\alpha}_{\beta}\, dx^{\beta} \otimes \frac{\partial}{\partial y^{\alpha}}$ on a $\dot{A} \circ \dot{B} = A^{\lambda}_{\alpha} B^{\alpha}_{\beta} dx^{\beta} \otimes \frac{\partial}{\partial y^{\alpha}}$.

Comme il est bien connu, si v est une connexion, la dérivée covariante associée à v est définie par $D_z w = (v w_*)_{(z)}$ où $z, w \in \chi(M)$ et w_* est l'application tangente à w.

<u>Définition 1.</u> <u>On appelle connexion généralisée sur M un tenseur $\tilde{v} \in \otimes^1_1(TM)$ vérifiant $J\tilde{v} = 0$.</u>
<u>$\tilde{H} = \operatorname{Ker} \tilde{v}$ est dit sous-espace horizontal.</u>
<u>Si $z, w \in \chi(M)$, on pose</u> $\tilde{D}_z w = (\tilde{v} w_*)_{(z)}$.

Puisque par définition $\operatorname{Im} \tilde{v} \subset VTM$, on a $\dim \tilde{H} \geq n$ ($n = \dim M$). Si $\tilde{H}$ est de dimension constante n et est transverse en tout point à VTM, alors $\tilde{v}$ définit naturellement une connexion au sens habituel. On vérifie sans difficultés que $\tilde{D}$ est un opérateur local et que $\tilde{D}_z w$ ne dépend que de la restriction de w aux courbes intégrales de z. On peut donc définir l'action de $\tilde{D}$ sur les champs le long d'une courbe.

<u>Définition 2.</u> <u>Une courbe γ sur M est dite géodésique de la connexion généralisée $\tilde{v}$ si</u> $\tilde{D}_{\gamma'} \gamma' = 0$.

Dans un système de coordonnées locales $\tilde{v}$ s'écrit sous forme matricielle

$$\tilde{v} = \begin{pmatrix} 0 & 0 \\ \tilde{\Gamma} & A \end{pmatrix}$$

où A et $\tilde{\Gamma}$ sont deux matrices (n,n). Les géodésiques vérifient le système

$$A^{\alpha}_{\beta}(x,\dot{x})\ddot{x}^{\beta} + \tilde{\Gamma}^{\alpha}_{\beta}(x,\dot{x})\dot{x}^{\beta} = 0$$

On pose $\dot{A} = \tilde{v}J$. $\dot{A}$ est sémi-basique et en particulier rg $\dot{A} \leqslant n$.

<u>Proposition 1. $\tilde{H}$ est de dimension n en tout point où il est transverse à VTM. De plus $\tilde{v}$ définit une connexion (au sens classique) si et seulement si rg $\dot{A} = n$ en tout point de TM</u> (c'est-à-dire $A_{(z)}$ est inversible pour tout $z \in TM$).

<u>Démonstration</u>. Supposons rg $\dot{A}_z = n$. Puisque $\dot{A} = J\tilde{v}$ et rg $J = n$ on a nécessairement rg $\tilde{v}_z = n$ et donc dim $\tilde{H}_z = n$. D'autre part rg $\dot{A}_z = n$ est équivalent à Im $\dot{A}_z$ = Ker $\dot{A}_z = V_zTM$. Soit $X \in \tilde{H}_z \cap V_zTM$. Puisque X est vertical il existe Y tel que $X = \dot{A}Y$ et puisque $X \in \tilde{H}_z$ on a $\tilde{v}\dot{A}Y = 0$. Or $\dot{A}$ est sémi-basique donc pour toute connexion v on a $v\dot{A} = \dot{A}$ d'où $\tilde{v}v\dot{A}Y = 0$ c'est-à-dire $\tilde{v}JF\dot{A}Y = 0$ et donc $\dot{A}F\dot{A}Y = 0$. Mais Ker $\dot{A}_z = V_zTM$ d'où $F\dot{A}Y$ est vertical; puisque par ailleurs il appartient à la distribution horizontale de v on a $F\dot{A}Y = 0$ d'où $\dot{A}Y = 0$, c'est-à-dire $X = 0$.

Réciproquement supposons $\tilde{H}_z \cap V_zTM \neq (0)$ et soit $Y \in \tilde{H}_z \cap V_zTM$, $Y \neq 0$. Puisque $Y \in V_zTM$ et $Y \neq 0$ il existe U non vertical tel que $Y = JU$. Or $\dot{A}Y = \tilde{v}JU = \tilde{v}Y = 0$ car $y \in \tilde{H}_z$; donc $U \in \text{Ker } \dot{A}_z$ et par conséquent Ker $\dot{A}_z \underset{\neq}{\supset} V_zTM$, c'est-à-dire rg $\dot{A}_z < n$.

cqfd.

<u>2.- CONSTRUCTION DE CONNEXIONS SINGULIERES</u>.

1. Soit ∇ une connexion linéaire (au sens habituel) sur TM, telle que $\nabla J = 0$. Alors $\tilde{v} = \nabla C$ est une connexion généralisée sur M dite projection de ∇. La proposition 1. exprime que $\tilde{v}$ est une connexion si et seulement si ∇ est régulière au sens de [4].

2. Soit $\omega \in \Lambda^1 M$ et $L \in \otimes^1_1(M)$; on définit $\omega L \in \Lambda^1 M$ par $(\omega L)(X) = \omega(LX)$, (c'est-à-dire $\omega L = i_L\omega$). On note $\underset{\sim}{\text{Ker}}\ L = \{\omega \in \Lambda^1 M : \omega L = 0\}$. Si $\dot{A}$ et $\dot{B}$ sont deux tenseurs sémi-basiques $\binom{1}{1}$ sur TM on a :

$$\underset{\sim}{\text{Ker}}\ \dot{A} \cap \underset{\sim}{\text{Ker}}\ \dot{B} \supset V\Lambda^1 M = \{\text{espace de 1-formes sémi-basiques}\} = $$
$$= \{\omega \in \Lambda^1 TM : \omega J = 0\}$$

et

$$\operatorname{Ker} \dot{A} \cap \operatorname{Ker} \dot{B} \supset VTM$$

Définition 3.- Deux tenseurs sémi-basiques $\dot{A}$ et $\dot{B}$ de type $\binom{1}{1}$ sur TM sont dits disjoints (respect. codisjoints) si $\operatorname{Ker} \dot{A} \cap \operatorname{Ker} \dot{B} = VTM$ (respect. : $\underset{\sim}{\operatorname{Ker}} \dot{A} \cap \underset{\sim}{\operatorname{Ker}} \dot{B} = V\Lambda^1 TM$).

La terminologie est justifiée par le fait que $\dot{A}$, $\dot{B}$ sont disjoints (respect. codisjoints) si et seulement si $\operatorname{Ker} A_{(z)} \cap \operatorname{Ker} B_{(z)} = \{0\}$ (respect. $\underset{\sim}{\operatorname{Ker}} A_{(z)} \cap \underset{\sim}{\operatorname{Ker}} B_{(z)} = \{0\}$).

Proposition 2.- Soit F la structure presque-complexe associée à une connexion sur M. Toute connexion généralisée s'écrit d'une manière unique sous la forme

$$\tilde{\nabla} = \dot{A}F + \dot{B}$$

où $\dot{A}$ et $\dot{B}$ sont deux tenseurs $\binom{1}{1}$ sémi-basiques sur TM. De plus $\operatorname{Dim} \tilde{H} = n$ si et seulement si $\dot{A}$ et $\dot{B}$ sont codisjoints.

Démonstration. Soit $\tilde{\nabla}$ une connexion généralisée. On pose $\dot{A} = \tilde{\nabla}J$ et $\dot{B} = \tilde{\nabla} - \dot{A}F$; on vérifie immédiatement que $\dot{A}$ et $\dot{B}$ sont sémi-basiques.

Réciproquement si $\tilde{\nabla} = \dot{A}F + \dot{B}$ on a $J\tilde{\nabla} = 0$ et donc $\tilde{\nabla}$ est une connexion généralisée.

D'autre part, soit $\tilde{\nabla} = \dot{A}F + \dot{B}$. On a évidemment $\underset{\sim}{\operatorname{Ker}} J \subset \underset{\sim}{\operatorname{Ker}} \tilde{\nabla}$ car $\operatorname{Im} \tilde{\nabla} \subset VTM$ et

$$\dim \tilde{H} = n \iff \operatorname{Im} \tilde{\nabla} = VTM \iff \underset{\sim}{\operatorname{Ker}} J = \underset{\sim}{\operatorname{Ker}} \tilde{\nabla}$$

Supposons $\dot{A}$ et $\dot{B}$ codisjoints et soit ω telle que $\omega\tilde{\nabla} = 0$. En multipliant à droite par J on obtient $\omega\dot{A} = 0$, d'où $\omega\dot{B} = 0$ et donc $\omega J = 0$. On a donc $\underset{\sim}{\operatorname{Ker}} \tilde{\nabla} = \underset{\sim}{\operatorname{Ker}} J$ et donc $\dim \tilde{H} = n$.

Réciproquement supposons que $\dim \tilde{H} = n$ et soit ω telle que $\omega\dot{A} = 0$ et $\omega\dot{B} = 0$. On a alors $\omega\tilde{\nabla} = \omega\dot{A}F + \omega\dot{B} = 0$, donc $\omega \in \underset{\sim}{\operatorname{Ker}} \tilde{\nabla}$. Puisque $\dim \tilde{H} = n$, $\omega \in \underset{\sim}{\operatorname{Ker}} J$ et donc $\dot{A}$ et $\dot{B}$ sont codisjoints.

cqfd.

3.- CONNEXIONS SINGULIERES ET CLASSE DE MASLOV.

Définition 4.- Soit Ω une structure symplectique sur TM. Une connexion généralisée $\tilde{\nabla}$ est dite lagrangienne si $\tilde{H}$ est un sous-fibré lagrangien de $(TTM,\Omega) \longrightarrow TM$.

Soient $\tilde{\nabla}$ et $\tilde{\nabla}'$ deux connexions généralisées lagrangiennes. On appelle classe du couple $(\tilde{\nabla},\tilde{\nabla}')$ la classe

$$M_\Omega(\tilde{v},\tilde{v}') = s^*\mathbb{M}(TTM,\hat{\tilde{H}},\hat{\tilde{H}}') \in H^1(M,\mathbb{Z})$$

où s est une section quelconque de $TTM \longrightarrow TM$ et $\mathbb{M}(TTM,\hat{\tilde{H}},\hat{\tilde{H}}')$ est la classe de Maslov -suivant la définition de Dazord (cf. [1])- du couple de fibrés lagragiens $\hat{\tilde{H}}$ et $\hat{\tilde{H}}'$.

<u>Exemples</u>. 1. Soit (M,g) une variété riemannienne, $\Omega = dd_J E$ la structure symplectique sur TM naturellement associée à g (cf. [3]). On a $\Omega(JX,JY) = 0$ $\forall X,Y \in TTM$. Ainsi J est une connexion généralisée lagrangienne; ici: $\tilde{H} = VTM$. Par conséquent à toute connexion généralisée lagrangienne $\tilde{v}$ sur (M,g) est associée canoniquement une classe de cohomologie entière

$$\alpha(\tilde{v}) = M_\Omega(J,\tilde{v})$$

Une construction analogue se fait évidemment sur les variétés finsleriennes.

2. Soit (M,ω) une variété symplectique. Le relèvement canonique ω^c de ω est une structure symplectique sur TM et l'on a $\omega^c(JX,JY) = 0$ $\forall X,Y \in TTM$. Par conséquent à toute connexion généralisée lagrangienne $\tilde{v}$ sur (M,ω) on peut associer une classe de cohomologie

$$\beta(\tilde{v}) = M_{\omega^c}(J.\tilde{v})$$

<u>Remarques</u>: 1. Si v est une connexion on a $\alpha(v) = 0$ (de même $\beta(v) = 0$) car H est transverse à VTM. La classe apparait ainsi liée à la dégénérescence de la connexion généralisée (c'est-à-dire aux intersections du vertical avec l'"horizontal").

2. La connexion de Levi-Civita d'une variété riemannienne est lagrangienne relativement à $\Omega = dd_J E$. De même toute connexion D linéaire symplectique sur une variété symplectique (c'est-à-dire telle que $D\omega = 0$ et ayant une torsion nulle) est lagrangienne.

4.- CONSTRUCTION DE CONNEXIONS SINGULIERES LAGRANGIENNES.

On introduit la notation suivante. Soit Ω une structure symplectique et L un tenseur $\binom{1}{1}$; on définit l'adjoint symplectique L^* de L par $\Omega(L^*X,Y) = \Omega(X,LY)$.

La propriété suivante est immédiate:

<u>Lemme</u>. <u>Une connexion généralisée $\tilde{v}$ est lagrangienne si et seulement si $\dim \tilde{H} = n$ et $\tilde{v}\tilde{v}^* = 0$.</u>

En effet si ces deux conditions sont réalisées on a Im $\tilde{\nabla}^* \subset \mathrm{Ker}\ \tilde{\nabla} = \tilde{H}$ et, puisque rg $\tilde{\nabla}$ = rg $\tilde{\nabla}^*$ on en déduit Im $\tilde{\nabla}^* = \tilde{H}$. D'autre part, pour tout $X,Y \in TTM$ on a $0 = \Omega(\tilde{\nabla}\tilde{\nabla}^*X,Y) = \Omega(\tilde{\nabla}^*X,\tilde{\nabla}^*Y)$; donc $\tilde{H}$ est lagrangienne. La réciproque est évidente.

Nous allons caractériser maintenant les connexions généralisées lagrangiennes dans le cas riemannien et dans le cas symplectique.

Théorème.- a) Soit (M,g) une variété riemannienne, F la structure presque complexe associée à la connexion de Levi-Civita. Toute connexion généralisée lagrangienne s'écrit d'une manière unique

$$\tilde{\nabla} = \dot{A}^*F + \dot{B}^* \quad (\dot{A}^* \text{ désignant l'adjoint de A relativement à } dd_JE)$$

où $\dot{A}$ et $\dot{B}$ sont deux tenseurs $\binom{1}{1}$ sémi-basiques sur TM disjoints et tels que $\dot{A}^*\circ\dot{B} = \dot{B}^*\circ\dot{A}$ (c'est-à-dire ${}^tA_{(z)}B_{(z)} = {}^tB_{(z)}A_{(z)}$ $\forall z \in TM$, où ${}^tA_{(z)}$ désigne le transposé de $A_{(z)}$ relativement à g) La classe de cohomologie sera notée $\alpha(\dot{A},\dot{B})$.

b) Soit (M,ω) une variété symplectique, F la structure presque-complexe associée à une connexion linéaire symplectique D sur M. Toute connexion généralisée lagrangienne s'écrit d'une manière unique

$$\tilde{\nabla} = \dot{A}^*F + \dot{B}^* \quad (\dot{A}^* \text{ désignant ici l'adjoint de A relativement à } \omega^c)$$

où $\dot{A}$ et $\dot{B}$ sont deux tenseurs $\binom{1}{1}$ sémi-basiques sur TM disjoints et tels que $\dot{A}^*\circ\dot{B} + \dot{B}^*\circ\dot{A} = 0$ (c'est-à-dire $A^*_{(z)}\circ B_{(z)} + B^*_{(z)}\circ A_{(z)} = 0$ $\forall z \in TM$, où $A^*_{(z)}$ désigne l'adjoint de $A_{(z)}$ relativement à ω).
De plus, si F et F' sont les structures presque-complexes associées à deux connexions symplectiques D et D' et si $\tilde{\nabla} = \dot{A}^*F + \dot{B}^*$ et $\tilde{\nabla}' = \dot{A}^*F' + \dot{B}^*$ on a: $\beta(\tilde{\nabla}) = \beta(\tilde{\nabla}')$. La classe de cohomologie de $\tilde{\nabla}$ peut donc être notée $\beta(\dot{A},\dot{B})$.

Démonstration: On voit immédiatement tout d'abord que $\dot{A}$ et $\dot{B}$ sont disjoints si et seulement si $\dot{A}^*$ et $\dot{B}^*$ sont codisjoints. Par conséquent toute connexion généralisée lagrangienne s'écrit nécessairement sous la forme $\tilde{\nabla} = \dot{A}^*F + \dot{B}^*$ avec $\dot{A}^*$ et $\dot{B}^*$ codisjoints, puisque dim $\tilde{H} = n$. (cf proposition 1).

La condition $\tilde{\nabla}\tilde{\nabla}^* = 0$ donne $\dot{A}^*FF^*\dot{A} + \dot{A}^*F\dot{B} + \dot{B}^*F^*\dot{A} + \dot{B}^*\dot{B} = 0$. Or $F^* = -F$ dans le cas riemannien et $F^* = F$ dans le cas symplectique. Puisque

$\dot{A}^*\dot{A} = 0$ et $\dot{B}^*\dot{B} = 0$, $\dot{A}$ et $\dot{B}$ étant sémi-basiques, on en déduit les relations supplementaires sur $\dot{A}$ et $\dot{B}$.

Pour la seconde partie il suffit de remarquer que si h et h' sont les projecteurs horizontaux de D et D', on a $h' = h + L$ où L est un tenseur sémi-basique. Du fait que h et h' sont lagrangiennes on déduit facilement que $i_L\omega^c = 0$ c'est-à-dire $L^* = -L$. F et F' sont donc liées par $F' = (I+L)F(I-L)$ (il suffit de vérifier que $F'J = h'$ et $F'h' = -J$). On en déduit que $\tilde{v}^{*\prime} = S\tilde{v}^*$ où $S = I + L$. Or $SS^* = I$, donc S est un symplectomorphisme.

Par conséquent

$$\mathbb{M}(TTM, VTM, \tilde{H}) = \mathbb{M}(TTM, S(VTM), S\tilde{H}) = \mathbb{M}(TTM, VTM, \tilde{H}')$$

c'est-à-dire $\beta(\tilde{v}) = \beta(\tilde{v}')$.

cqfd.

Exemples.

D'après ce théorème, on peut construire sur une variété riemannienne une classe de cohomologie entière dans les cas suivants:

1. Par la donnée de deux tenseurs A, $B \in \otimes^1_1(M)$ tels que $\mathrm{Ker}A \cap \mathrm{Ker}B = \{0\}$ et ${}^tAB = {}^tBA$ (on prend $\dot{A} = A^v$ (relèvement vertical de A) et $\dot{B} = B^v$).

2. Par la donnée d'un tenseur symétrique $A \in \otimes^1_1(M)$ et d'une fonction $f \in C^\infty(M)$ telle que $f|_{\Sigma_A} \neq 0$ où $\Sigma_A = \{x \in M : \text{dét}\, A_x = 0\}$ (On prend $B = fI$). La classe est notée $\alpha(A,f)$.

3. Par la donnée de deux fonctions f, g non simultanément nulles (On prend $A = fI$ et $B = gI$). La classe est notée $\alpha(f,g)$.

4. Soit $M \longrightarrow M'$ une immersion isométrique d'une sous-variété de codimension 2 dans une variété à courbure constante. Si ξ et η sont deux champs normaux indépendants soient $A = K_\xi$ et $B = K_\eta$ (endomorphismes de Weingarten). La connexion généralisée associée à A et B est lagrangienne si et seulement si le fibré normal est plat et l'indice de nullité relative est nul. On peut donc dans ce cas construire une classe de cohomologie entière et l'on montre qu'elle ne dépend pas du choix de ξ et η. (cf. [7] proposition p. 46)

Les propriétés principales de la classe $\alpha(\dot{A},\dot{B})$ sont données en [6]. En particulier: $\alpha(\dot{A},\dot{B}) = \alpha(\dot{B},-\dot{A})$ et $\alpha(\dot{A},f) = 0$ si $f \neq 0$. Dans cette note on montre aussi qu'à toute connexion lagrangienne généralisée $\tilde{v}$ est associée canoniquement une connexion lagrangienne généralisée $\tilde{\nabla}_1$ dite normalisée, ayant même espace horizontal que $\tilde{\nabla}$ et donc mêmes géodésiques et même classe de cohomologie. Les connexions ainsi normalisées forment un groupe N pour une certaine loi de composition et l'application

$$\begin{aligned} \alpha: N &\longrightarrow H^1(M,\mathbb{Z}) \text{ est un homomorphisme.} \\ \tilde{\nabla} &\longrightarrow \alpha(\tilde{\nabla}) \end{aligned}$$

Remarquons enfin que l'on peut définir la courbure d'une connexion lagrangienne généralisée par le tenseur sémi-basique $\binom{1}{2}$ sur TM

$$\tilde{R} = -\frac{1}{2}\tilde{\nabla}\,[\tilde{\nabla}^*,\tilde{\nabla}^*]$$

Dans le cas d'une connexion ce tenseur coïncide avec la courbure habituelle (cf. [3]).

Proposition 3.- $\tilde{H}$ définit un feuilletage lagrangien si et seulement si $\tilde{R} = 0$.

Démonstration. Il suffit de remarquer que $\tilde{R}(X,Y) = -\tilde{\nabla}\,[\tilde{\nabla}^*X,\tilde{\nabla}^*Y]$.

5.- CALCUL DE L'INDICE D'UNE COURBE ASSOCIÉ A LA CLASSE $\alpha(\dot{A},\dot{B})$.

Proposition 4.- Soit $\Sigma_A = \pi\{z \in TM : \text{rg } A_{(z)} < \text{Sup rg } A_{(z)}\}$ et γ une courbe de M qui coupe transversalement Σ_A aux points $t_1,\dots,t_m$. On a

$$\int_\gamma \alpha(\dot{A},\dot{B}) = \sum_{i=1}^{m}\left[\left(\text{ind}\left({}^tA_{(z)}B_{(z)}\right)_{\gamma(t_i+\varepsilon)} - \text{ind}\left({}^tA_{(z)}B_{(z)}\right)_{\gamma(t_i-\varepsilon)}\right]\right.$$

(Remarque: $\forall z \in TM$ ${}^tA_{(z)}B_{(z)}$ est un tenseur symétrique sur M (cf. théorème) ind désigne l'indice de la forme quadratique associée).

En particulier

$$\int_\gamma \alpha(f,g) = -\frac{n}{2}\sum_{i=1}^{n}\left[\text{sign}(fg)_{\gamma(t_i+\varepsilon)} - \text{sign}(fg)_{\gamma(t_i-\varepsilon)}\right] =$$

$$= -n\sum_{i=1}^{m}\text{sign}\left(f\frac{dg}{dt}\right)_{\gamma(t_i)}$$

t_i étant les points où γ intersecte $\Sigma_f = \{x \in M : f(x) = 0\}$.

Démonstration. Soit $\tilde{\gamma}$ une courbe sur TM; on a la formulle (cf. [1])

$$\int_{\gamma} \mathbb{M}\ (TTM,VTM,\tilde{H}) = \sum_{i=1}^{m} \mathrm{ind}\left\{ Q^{(VTM,\tilde{H},L)}_{\tilde{\gamma}(t_i+\varepsilon)} - Q^{(VTM,\tilde{H},L)}_{\tilde{\gamma}(t_i-\varepsilon)} \right\}$$

où L est un sous-fibré lagrangien arbitraire de TTM $\longrightarrow$ TM, t_i les points où $\tilde{\gamma}$ rencontre $\Sigma = \{z \in TM : \dim(VTM \cap \tilde{H})_z > \mathrm{Inf}\ \dim(VTM \cap \tilde{H})_z\}$ et $Q^{(VTM,\tilde{H},L)}$ la forme quadratique sur $\tilde{H}$ définie par $Q(x) = \Omega(X, Pr_{V/\!/L}\, X)$, $Pr_{V/\!/L}$ étant la projection sur VTM parallélement à L.

Or la distribution horizontale H de la connexion de Levi-Civita est lagrangienne; on peut donc prendre L = H. D'autre part, d'après la démonstration de la proposition 1. on a $\dim(VTM \cap \tilde{H}) = n - \mathrm{rg}\ \dot{A}$ et $\mathrm{ind}\ Q_z = \mathrm{ind}\ {}^tA_{(z)}B_{(z)}$. D'où la formulle.

<u>Corollaire</u>. a) <u>$\alpha(A,f)$ ne depend que de A et du signe de f sur</u> Σ_A. <u>En particulier $\alpha(f,g)$ ne depend que du signe de f sur</u> Σ_g <u>et du signe de g sur</u> Σ_f.

b) <u>$\alpha(A,f) = 0$ si</u> Σ_A <u>est connexe</u>.

a) est évidente. Pour montrer b) il suffit de constater que d'après a) $\alpha(A,f) = \alpha(A,\tilde{f})$ pour toute fonction $\tilde{f}$ qui coïncide avec f sur Σ_A. Puisque $f|_{\Sigma_A} \neq 0$ si Σ_A est connexe f est de signe constant sur Σ_A. On peut choisir $\tilde{f}$ de signe constant sur M. D'après ([6] proposition 2.) $\alpha(A,\tilde{f}) = 0$, donc $\alpha(A,f) = 0$.

<u>Exemple d'une classe non triviale</u>.

<u>Proposition</u> 5. <u>Soit</u> $h: M \longrightarrow \mathbb{R}^2$ <u>une submersion</u> ($h = (f,g)$) <u>et</u> $N = h^{-1}(0)$. <u>Alors $\alpha(f,g)$ est un élément non trivial de $H^1(M-N,\mathbb{Z})$</u>.

<u>Démonstration</u>. Au voisinage d'un point de $h^{-1}(0)$ on peut choisir des coordonnées locales telles que $h(x_1,\dots,x_n) = (x_1,x_2)$: ainsi $f = x_1$ et $g = x_2$. Soit γ définie par $x_1^2 + x_2^2 = 1$, $x_3 = x_4 = \dots = x_n = 0$. Σ_f est réduit aux points $(0,-1,0,\dots,0)$ et $(0,1,0,\dots,0)$. En utilisant la formulle de la proposition (4) on trouve $(f,g) = 2n$.

6. REPRESENTANT DE MORVAN-DAZORD DE $\alpha(\dot{A},\dot{B})$.

J. M. Morvan a donné le premier un représentant métrique de la classe de Maslov d'une sous-variété lagrangienne de $\mathbb{R}^{2n}$; Dazord a ensuite généralisé ce résultat au cas de deux fibrations lagrangiennes (cf. [10] et [2]). A l'aide de ces résultats nous allons construire maintenant un représentant de la classe $\alpha(\dot{A},\dot{B})$. Introduisons d'abord la notation suivante. Soit $\dot{A}$ un tenseur $\binom{1}{1}$ sémi-basique sur TM. F la structure presque-complexe associée à une connexion. $\mathrm{Tr}(F\dot{A}) \in C^\infty(TM)$ (Tr désignant la trace) est indépendante du choix de la connexion et est notée $\dot{\mathrm{T}}\mathrm{r}\, A$. On a $(\dot{\mathrm{T}}\mathrm{r}\, A)_z = \mathrm{Tr}\, A_{(z)}$, c'est-à-dire si $\dot{A} = A^\beta_\alpha(x,y)\, dx^\alpha \otimes \frac{\partial}{\partial y^\beta}$
$\dot{\mathrm{T}}\mathrm{r}\, A = A^\alpha_\alpha(x,y)$.

Soient L et M deux tenseurs $\binom{1}{1}$ sémi-basiques sur TM ; supposons M de rang n (c'est-à-dire $M_{(z)}$ est invertible pour tout $z \in TM$). Soit $M^{(-1)}$ le tenseur $\binom{1}{1}$ sémi-basique sur TM définie par $(M^{(-1)})_{(z)} = M^{-1}_{(z)}$ (cf. [6]). On a $\dot{\mathrm{T}}\mathrm{r}(L\circ M^{(-1)}) = \dot{\mathrm{T}}\mathrm{r}(M^{(-1)}\circ L)$; on peut donc noter cette expression par $\dot{\mathrm{T}}\mathrm{r}\, \frac{L}{M}$.

Proposition 6.- Le représentant de Morvan-Dazord de la classe $\alpha(\dot{A},\dot{B})$ est

$$\alpha_{MD}(\dot{A},\dot{B})(z) = \frac{1}{\pi}\,\dot{\mathrm{T}}\mathrm{r}\left(\frac{\dot{A}^*\circ\nabla_{s*z}\dot{B} - \dot{B}^*\circ\nabla_{s*z}\dot{A}}{\dot{A}^*\circ\dot{A} + \dot{B}^*\circ\dot{B}}\right)_s$$

où $z \in TM$, $s \in \chi(M)$ et ∇ est le relèvement de Dombrowski sur TM de la connexion de Levi-Civita (ou le relèvement de Cartan dans le cas finslerien : cf. [4]).

En particulier, si $\dot{A} = A^V$ et $\dot{B} = B^V$ avec $A, B \in \otimes^1_1(M)$

$$\alpha_{MD}(A,B)(X) = \frac{1}{\pi}\,\mathrm{Tr}\left(\frac{{}^tAD_XB - {}^tBD_XA}{{}^tAA + {}^tBB}\right)$$

(où $X \in TM$ et D est la connexion de Levi-Civita) et

$$\alpha_{MD}(f,g) = \frac{n}{\pi}\left(\frac{fdg - gdf}{f^2 + g^2}\right)$$

Démonstration. On a (cf. [2])

$$\mathbb{M}\,(TTM,VTM,\tilde{H}) = \frac{1}{\pi}\sum_{j=1}^{n}\left(<\nabla e^{\tilde{H}}_j, Fe^{\tilde{H}}_j> - <\nabla e^V_j, Fe^V_j>\right)$$

où $\{e^{\tilde{H}}_j\}$ et $\{e^V_j\}$ sont deux bases orthonormées respectivement de $\tilde{H}$ et de

VTM.

Soit $U = -\tilde{\nabla}F\tilde{\nabla}^*$. Pour tout $z \in TM$, $U_{(z)}$ est symétrique et défini positif (cf. [6]); il existe donc une base orthonormée de vecteurs horizontaux (pour la connexion de Levi-Civita) $Y_1, \ldots, Y_n$ tels que $UY_i = \lambda_i JY_i$ avec $\lambda_i > 0$ (il suffit de prendre le relèvement horizontal d'une base orthonormée $\{\varepsilon_i\}$ de $T_{\pi(z)}M$ telle que $U_{(z)}\varepsilon_i = \lambda_i \varepsilon_i$). On pose alors

$$e_i^{\tilde{H}} = \frac{\tilde{\nabla}^* Y_i}{\sqrt{\lambda_i}} \qquad \text{et} \qquad e_i^V = JY_i$$

Tenant compte du fait que $U = -(\dot{A}^* \circ \dot{A} + \dot{B}^* \circ \dot{B})$ et que $\nabla J = 0$ et $\nabla F = 0$ (cf. [4]) un simple calcul donne la formule cherchée.

Corollaire .- $\alpha(\dot{A},\dot{B}) = -\alpha(\dot{B},\dot{A})$

(Cette propriété n'était pas évidente sur les formules de la proposition 4.)

Le calcul de la classe $\alpha(\dot{A},\dot{B})$ peut se ramener à celui de la classe de deux fonctions $\alpha(f,g)$. Soit $\dot{A}$ un tenseur $\binom{1}{1}$ sémi-basique sur TM. On définit $\det \dot{A} \in C^\infty(TM)$ par $(\det \dot{A})_z = \det A_{(z)}$.

En utilisant le complexifié de TTM, on pose $L = \dot{A} + i\dot{B}$. On a

Proposition 7.

$$\alpha(\dot{A},\dot{B}) = \frac{1}{n}\,\alpha(\mathrm{Re}\,\det L,\ \mathrm{Im}\,\det L)$$

(où Re et Im désignent la partie réelle et la partie imaginaire)

Pour la démonstration: cf. [7] et [8].

Corollaire .- Si A est symétrique et B antisymétrique, alors $\alpha(A,B) = 0$.

En effet, dans ce cas L est hermitienne et donc $\det L$ est réel.

7.- APPLICATIONS:

On se limitera ici à quelques résultats simples. Pour les démonstrations cf. [8].

1) Equations différentielles sur S^1.

Soit $\gamma: S^1 \longrightarrow \mathbb{R}$ une courbe de classe C^1. On a $\alpha(1,\gamma') = 0$ (cf. [6] proposition 2) donc, d'après la proposition 4, $\sum_i \mathrm{sign}\,\gamma''(\theta_i) = 0$ (θ_i tels que $\gamma'(\theta_i) = 0$). C'est-à-dire, comme il est bien connu: le nombre des maximums du graph de γ est égal au nombre de minimums.

Plus généralement α permet d'exprimer le nombre des maximums moins le nombre des minimums pour une courbe avec des discontinuités.

Soit l'équation différentielle sur I $g(t)\dot{x}(t) = f(t)$, $U = \{t\in I, g(t) \neq 0\}$ et (U_k) les composantes connexes de U. On appelle solution généralisée une application $x: U \longrightarrow \mathbb{R}$ telle que $x|_{U_k}$ soit une solution maximale sur U_k.

Proposition 8. Soient $f, g \in C^\infty(S^1)$, $\tilde{\nabla}$ la connexion généralisée associée à f, g et $\bar{\gamma}$ une solution généralisée de l'équation des relevées sur TS^1 des géodésiques de $\tilde{\nabla}$. Alors

$$\alpha(\tilde{\nabla}) = \text{(Nb max - Nb min) du graph de } \bar{\gamma} = $$
$$= 2\,\text{Nb de tours effectués par la tangente à } \bar{\gamma}).$$

2) Indice d'un champ de vecteurs dans le plan.

Proposition 9.- Soit $X = f(x,y)\frac{\partial}{\partial x} + g(x,y)\frac{\partial}{\partial y}$ un champ sur $\mathbb{R}^2$ tel que $X_m = 0$ et soit $J(X,m)$ l'indice de X en m. On a:

$$J(X,m) = \frac{1}{4}\int_\gamma \alpha(f,g)$$

où γ est un lacet entourant m et n'entourant pas d'autres points singuliers de X.

3) Courbure algébrique totale d'une courbe plane.

Proposition 10.-Soit $\gamma: I \longrightarrow \mathbb{R}^2$ une courbe de classe C^2, k sa courbure et $K = \int_I k\,dt$ sa courbure algébrique totale.
Soit a un vecteur constant et $f = \langle a,\vec{t}\rangle$, $g = \langle a,\vec{n}\rangle$ ($\vec{t}$, $\vec{n}$ étant le repère de Frenet). Alors

$$K = \pi\int_\gamma \alpha(f,g)$$

4) Possibles applications à la physique.

Les équations des géodésiques d'une connexion généralisée peuvent être considerées comme définissant l'état limite d'un système mécanique dont le champ des forces externes est très grand au voisinage de certains points (i.e. là où A dégénère). Conformement à la proposition 8. α pourrait évaluer le nombre des maximums moins le nombre des minimums du graph de $\bar{\gamma}$ en se limitant aux pics qui " ne sont pas trop élevés".

Un exemple de tels systèmes est donné par Takens ,[11],dans l'étude des réseaux de circuits électriques non linéaires.

Signalons enfin que la classe $\alpha(f,g)$ est sans doute liée à la notion de "Kink number" des solutions de l'équation de KdV (cf. [9]).

BIBLIOGRAPHIE.

[1] P. Dazord Invariants homotopiques attachés aux fibrés symplectiques. Ann. Inst. Fourier 1979 XIX, 2.

[2] P. Dazord Sur la géometrie des sous-fibrés et des feuilletages lagrangiens. Ann. Ec. Norm. Sup. 1982.

[3] J Grifone Structure presque tangente et connexions I. Ann. Inst. Fourier 1972 XXII, 1.

[4] J. Grifone Structure presque tangente et connexions II. Ann. Inst. Fourier 1972, XXII, 3.

[5] J. Grifone Connexions singulières et classe de Maslov. CRAS (295) 1982, p.139.

[6] J. Grifone Le groupe des connexions singulières normalisés. CRAS (295) 1982, p.273.

[7] J. Grifone - E. Hassan. Calcul de la classe de cohomologie d'une connexion singulière lagrangienne et applications. CRAS (295) 1982 p.543.

[8] E. Hassan Connexions singulières et classe de Maslov. Thèse III cycle . Toulouse 1982.

[9] R. Hermann Toda lattices, cosymplectic manifolds, Bäcklund transformations and Kinks. Interdisc. Math. Vol XV, part A.

[10] J. M. Morvan Classe de Maslov d'une immersion lagrangienne et minimalité. CRAS (292) 1981 p.633.

[11] F. Takens Constrained equations; a study of implicit differential equations and their discontinuous solutions. Math. Inst. St. Univ. Cromingen. The Netherlands.

SUR LA COHOMOLOGIE DES SYSTÈMES D'ÉQUATIONS DIFFÉRENTIELLES ET DES PSEUDOGROUPES DE LIE

A. Kumpera
Instituto de Matemática
Universidade Estadual de Campinas
13100 Campinas SP. Brasil

1.- Introduction.

En reprenant les techniques développées dans [2], nous introduisons sur les variétés Grassmanniennes un calcul différentiel extérieur tangent aux éléments de contact. Ce calcul jouit de toutes les propriétés opérationnelles habituelles, est localement exact (lemme de Poincaré) et incorpore le mécanisme du prolongement, par quoi sa restriction à une structure différentielle tient compte des éléments infinitésimaux d'ordre supérieur associés à cette structure. Le complexe de *de Rham* correspondant détermine alors des classes de cohomologie (locales et globales) qui reproduisent, par restriction du calcul aux variétés Ehresmanniennes (jets) et à un ordre fixé k, les classes partielles de Molino ([5],[6]). En particulier, on retrouve par des méthodes bien aisées et directes les classes de Bott-Haefliger d'un feuilletage aussi bien que les classes de cohomologie introduites dans [3], [4] et [8]. D'autre part, l'action prolongée de difféomorphismes locaux et de champs de vecteurs permet de définir une notion d'invariance pour les champs et les formes différentielles de ce calcul. La restriction du complexe de *de Rham* aux formes invariantes par une famille quelconque Γ de transformations finies ou infinitésimales locales produit un nouveau complexe dont la cohomologie est associée à Γ. En particulier, lorsque Γ est un pseudogroupe de transformations, on peut ainsi définir la cohomologie de Γ qui étend celle des groupes de Lie. Enfin, si l'on rassemble les deux procédés, on arrive à la cohomologie équivariante qui semble jouer un rôle important dans l'étude des équations différentielles invariantes par l'action de pseudogroupes de Lie (théorie de Lie-Vessiot).

Je tiens à remercier mon ami et collègue Luiz A. B. San Martin pour l'aide précieuse.

2.- Calcul covariant dans les variétés Ehresmanniennes.

Soient V et W deux variétés différentiables séparées de classe C^∞ et de dimensions respectives n et m, $\pi: V \to W$ une fibration (= surmersion) et J_kV la variété des k-jets de sections locales de π. Avec ces notations, $J_0V = V$, $J_{-1}V = W$ et si l'on indique par $\rho_{h,k}: J_kV \longrightarrow J_hV$, $h \leq k$, la projection canonique, alors $\rho_{-1,k} = \alpha_k : J_kV \longrightarrow W$ est la projection *source* et $\rho_{0,k} = \beta_k : J_kV \longrightarrow V$ la projection *but*. Soit $H_k: TJ_kV \longrightarrow TJ_{k-1}V$, $k \geq 1$, le morphisme de fibrés vectoriels défi ni par le *relèvement holonôme* (cf. [2]) c'est-à-dire, $H_k v = (j_{k-1}\sigma)_* \circ (\alpha_k)_* v$ où $v \in T_X J_k V$, $X = j_k\sigma(x)$ et $j_{k-1}\sigma : y \longmapsto j_{k-1}\sigma(y)$ est le flot holonôme associé à σ . On pose $H_0 = \pi_*$ et $H_{-1} = Id$. Chaque sous-espace vectoriel $im(H_k)_X$, $X \in J_k V$,est un élément de contact horizontal (i.e., transverse aux fibres de α_{k-1}) et l'application $X \longmapsto im(H_k)_X$ peut être envisagée comme une *connexion à paramètres* au dessus de la variété W , ce qui permet de développer un calcul covariant.

Soit Λ_k^* le module des formes différentielles extérieures α_k-semi-basiques de la variété $J_k V$ c'est-à-dire, des formes ω telles que $i(\xi)\omega = 0$ pour tout champ α_k-vertical ξ . En choisissant des coordonnées locales (x^i) de W et en les remontant à $J_k V$, les éléments de Λ_k^* s'écrivent localement par

(2.1) $$\omega = \Sigma\, f_{i_1 \ldots i_r}\, dx^{i_1} \wedge \ldots \wedge dx^{i_r}$$

où les coefficients $f_{i_1 \ldots i_r}$ sont des fonctions C^∞ sur $J_k V$. On définit une différentielle

$$d_k : \Lambda_k^* \longrightarrow \Lambda_{k+1}^*$$

en posant $d_k \omega = d\omega \circ \Lambda H_{k+1}$ c'est-à-dire, d_k est la différentielle covariante par rapport à la *connexion* H_{k+1} . On constate immédiatement que $d_{k+1} \circ d_k = 0$ et que d_k est une dérivation de degré 1 de Λ_k^* dans Λ_{k+1}^* relative au morphisme $\rho_{k,k+1}^*$. De plus, si l'on reprend l'expression locale (2.1) , on voit facilement (en abrégeant d_k par d) que $dx^i = dx^i$, $d\omega = \Sigma\, df_{i_1 \ldots i_r} \wedge dx^{i_1} \wedge \ldots \wedge dx^{i_r}$ et, pour toute fonction $f \in \Lambda_k^0$, $df = \Sigma\, \partial_i f\, dx^i$ où

$$\partial_i f = \partial f / \partial x^i + \sum_{|\alpha| \leq k} (\partial f / \partial y_\alpha^\lambda)\, y_{\alpha+1_i}^\lambda$$

est la dérivée *totale* de la fonction f par rapport au champ $\partial / \partial x^i$, $\alpha = (\alpha_1, \ldots, \alpha_m)$ est un multi-indice, $\alpha + 1_i = (\alpha_1, \ldots, \alpha_i + 1, \ldots, \alpha_m)$, $(x^i, y_\alpha^\lambda)_{|\alpha| \leq k+1}$ est le système de coordonnées locales de $J_{k+1} V$ associé aux coordonnées (x^i, y^λ) de V et $y_0^\lambda = y^\lambda$. On vérifie également, par un calcul local, que la condition $d_k f = 0$ est équivalente à la constance de la fonction f sur chaque composante connexe de $J_k V$. Du reste, cette propriété traduit le fait que le *système dérivé* de la structure de contact canonique de $J_{k+1} V$ est égal à la structure de contact canonique de $J_k V$ remontée à $J_{k+1} V$ (cf. [3]).

Puisque l'application $\rho_{k,k+1}^* : \Lambda_k^* \longrightarrow \Lambda_{k+1}^*$ est injective et que le diagramme

$$\begin{array}{ccc} \Lambda_{k+1}^* & \xrightarrow{d_{k+1}} & \Lambda_{k+2}^* \\ {\scriptstyle \rho_{k,k+1}^*} \uparrow & & \uparrow {\scriptstyle \rho_{k+1,k+2}^*} \\ \Lambda_k^* & \xrightarrow{d_k} & \Lambda_{k+1}^* \end{array}$$

est commutatif, la famille des différentielles d_k détermine une différentielle unique $d : \Lambda^* \longrightarrow \Lambda^*$ sur la limite inductive Λ^* de la famille (Λ_k^*, ρ^*) ; c'est une dérivation de degré 1 et de carré nul. La condition $df = 0$, $f \in \Lambda^0$, est équivalente à la constance locale de f et $d(\Lambda^m) = 0$.

Lemme de Poincaré. Le complexe $0 \longrightarrow R \longrightarrow \Lambda^0 \xrightarrow{d} \Lambda^1 \xrightarrow{d} \cdots \xrightarrow{d} \Lambda^m$ est localement exact (i.e., au niveau des germes).

Pour la démonstration, on renvoie le lecteur à [9], lemme A.4, en remarquant que Λ^* s'identifie à l'algèbre $\Sigma\, \Phi^{0,s}$ de [9] bien que les définitions soient sensiblement différentes. La partie finale $\Lambda^{m-1} \xrightarrow{d} \Lambda^m \xrightarrow{d} 0$ n'est pas localement exacte et peut être augmentée à droite par le complexe d'*Euler-Lagrange* introduit par Tulczyjew dans [9] .

Remanions maintenant les définitions de sorte à les adapter à la discussion du §3 . Soit X_k le Λ_k^0-module de tous les relèvements différentiables $\xi\colon J_kV \longrightarrow TW$ de α_k et X la limite inductive de la famille $(X_k, {}^t\rho)$ où ${}^t\rho(\xi) = \xi \circ \rho_{k,k+1}$. Le Λ^0-module X est une R-algèbre de Lie fléchée ([2],[7]). En effet, on peut envisager les éléments de X_k comme des opérateurs différentiels (non-linéaires) d'ordre k de $\Gamma(\pi)$ dans $\Gamma(TW)$ et le crochet $[\xi,\eta]$ de deux éléments de X_k , défini par la formule $[\xi,\eta](\sigma) = [\xi(\sigma),\eta(\sigma)]$, est alors un élément de X_{k+1} . De plus, puisque cette définition est compatible avec ${}^t\rho$, l'opération ci-dessus détermine un crochet d'algèbre de Lie sur X . De façon plus naïve, si $X = j_{k+1}\sigma(x) \in J_{k+1}V$, on pose $[\xi,\eta]_X = [\xi\circ j_k\sigma, \eta\circ j_k\sigma]_X$, le deuxième membre étant entièrement déterminé car X caractérise $j_1(j_k\sigma)x$. Soit

$$\lambda_k\colon J_kV \underset{\alpha}{\times} TW \longrightarrow TJ_{k-1}V$$

le relèvement holonôme ([2]). À tout $\xi \in X_k$ est associée une variation infinitésimale $\tilde{\xi}\colon J_kV \longrightarrow TJ_{k-1}V$ de $\rho_{k-1,k}$ définie par $\tilde{\xi}(X) = \lambda_k(X,\xi(X))$. On définit la *dérivée formelle* (ou *totale*) $\partial_\xi\colon \Lambda^0 \longrightarrow \Lambda^0$, $\xi \in X$, en posant $\partial_\xi f(X) = \tilde{\xi}(X)\cdot f$ où $\xi \in X_k$ et $f \in \Lambda_{k-1}^0$. La représentation $\partial\colon X \longrightarrow \mathrm{Der}\,\Lambda^0$ est le fléchement de X car $[\xi,f\eta] = f[\xi,\eta] + (\partial_\xi f)\eta$. Considérons ensuite le Λ_k^0-module $\tilde{\Lambda}_k^*$, ensemble des relèvements différentiables $\omega\colon J_kV \longrightarrow \Lambda^*W$ de α_k . Par définition, ω_X est une forme extérieure sur $T_{\alpha(X)}W$ ce qui montre immédiatement que $\tilde{\Lambda}_k^*$ s'identifie canoniquement à Λ_k^* . De plus, la différentielle $d\colon \Lambda^* \longrightarrow \Lambda^*$ peut être re-définie par la formule habituelle

$$\begin{aligned} d\omega(\xi_1,\dots,\xi_{r+1}) &= \Sigma\,(-1)^{i+1}\,\partial_{\xi_i}\omega(\xi_1,\dots,\hat{\xi}_i,\dots,\xi_{r+1}) \; + \\ &+ \sum_{i<j}(-1)^{i+j}\,\omega([\xi_i,\xi_j]\,\xi_1,\dots,\hat{\xi}_i,\dots,\hat{\xi}_j,\dots,\xi_{r+1}) \; . \end{aligned} \tag{2.2}$$

3.- Calcul covariant dans les variétés Grassmanniennes.

Soit V une variété différentiable (C^∞) de dimension finie et G_kV la Grassmannienne des éléments de contact d'ordre k et de dimension p de V c'est-à-dire, l'ensemble des classes d'équivalence de germes de sous-variétés ν de dimension p par la relation de tangence d'ordre k . Dans tout ce qui suit, l'entier p demeurera fixé. Nous indiquons par $\nu_k x$ l'élément de contact d'ordre k défini, au point

x , par la sous-variété ν , par $\rho_{hk}: G_kV \longrightarrow G_hV$, $h \le k$, la projection canonique et par E le fibré vectoriel canonique de base G_1V , les points de la base étant les éléments de contact linéaires de dimension p de la variété V et la fibre E_X étant l'espace vectoriel X . On introduit des coordonnées dans G_kV en l'identifiant localement à une variété Ehresmannienne comme suit: Étant donné un élément de contact ν_kx , il existe toujours une fibration locale $\pi: U \to W$ *greffée* à V au voisinage de x (i.e., U est un voisinage ouvert de x et W une variété) telle que ν_kx soit un élément de contact strictement transverse à π . On l'obtient, par exemple, en prenant pour U le domaine d'un système de coordonnées locales arbitraire défini au voisinage de x et pour π une quelconque des projections sur un espace de p coordonnées dont la restriction à $\nu_1x = T_x\nu$, élément de contact d'ordre un induit par ν_kx , soit bijective. Ceci étant, J_kU est une variété munie de coordonnées et s'identifie à un ouvert de G_kV .

Un champ de vecteurs d'ordre k sur V est une application différentiable $\xi: G_kV \longrightarrow E$ compatible avec ρ_{1k} c'est-à-dire, vérifiant $\xi(X) \in E_{\rho_{1k}X}$. Une forme différentielle d'ordre k est une application différentiable $\omega: G_kV \longrightarrow \Lambda E^*$ compatible avec ρ_{1k} . Étant donnés deux champs de vecteurs ξ,η d'ordre k , on définit le champ $[\xi,\eta]$ d'ordre $k+1$ par $[\xi,\eta](X) = [\xi_\nu,\eta_\nu]_x$ où $X \in G_{k+1}V$, $x = \rho_0X$, $\rho_0: G_{k+1}V \longrightarrow V$ est la projection naturelle, ν est un germe de sous-variété représentant X , $\xi_\nu = \xi \circ \nu_k$ et finalement $\nu_k: \nu \longrightarrow G_kV$ est l'application qui à chaque $y \in \nu$ fait correspondre l'élément de contact d'ordre k de la sous-variété ν en le point y . En se plaçant dans le contexte restreint de l'Ehresmannienne J_kV des jets de sections d'une fibration, on retrouve bien le crochet défini au §2 . Il est facile de vérifier que cette définition est indépendante du représentant ν . Soit χ_k le module (sur l'anneau F_k des fonctions différentiables de G_kV) de tous les champs d'ordre k et χ la limite inductive de la famille $(\chi_k, {}^t\rho_{hk})$ où ${}^t\rho_{hk}$: $\xi \in \chi_h \longrightarrow \xi \circ \rho_{hk} \in \chi_k$. Le crochet [,] définit alors une structure d'algèbre de Lie réelle sur le module χ , cette algèbre étant fléchée par la dérivée formelle $\partial: \chi \longrightarrow \mathrm{Der}\, F$ (définie par un processus de relèvement holonôme entièrement analogue à celui indiqué au §2) c'est-à-dire, on a la relation $[\xi,f\eta] = f[\xi,\eta] + (\partial_\xi f)\eta$ pour tout $f \in F = \varprojlim F_k$. On définit ensuite la différentielle extérieure d'une forme ω d'ordre k par la formule (2.2) , l'ordre de $d\omega$ étant égal à $k+1$. En indiquant par Λ_k^* le module des formes différentielles d'ordre k et par Λ^* la limite inductive de la famille $(\Lambda_k^*,\rho_{hk}^*)$, on obtient un opérateur différentiel $d: \Lambda^* \longrightarrow \Lambda^*$ de carré nul qui opère en tant que dérivation de degré un i.e., $d(\omega\wedge\mu) = d\omega \wedge \mu + (-1)^{\deg \omega}\, \omega \wedge d\mu$.

Lemme de Poincaré. Le complexe $0 \longrightarrow R \xrightarrow{d} \Lambda^0 \xrightarrow{d} \Lambda^1 \xrightarrow{d} \cdots\cdots \xrightarrow{d} \Lambda^p$ est localement exact.

Démonstration. Prenons une fibration $\pi: U \to W$, $U \subset V$, greffée à V et considérons la carte affine J_kU de G_kV . On remarque, et ceci pour une fibration quelcon que π , que le module $\mathcal{X}_k$, $k \geq 1$, des relèvements $\xi: J_k U \longrightarrow TW$ de α_k (cf. §2) s'identifie, à l'aide de π_* , au module des relèvements $\bar{\xi}: J_k U \longrightarrow \bar{E}$ de ρ_{1k} , où $\bar{E}$ est la restriction de E à la Grassmannienne des éléments de contact transverses $\bar{G}_1 U = J_1 U$, ce dernier module n'étant autre que $\mathcal{X}_k | J_k U$ (l'espace $\mathcal{X}_k$ étant celui défini ci-avant au niveau des Grassmanniennes). Les définitions montrent de plus que cette identification est compatible avec les crochets et les flèchements. De même, le module $\tilde{\Lambda}_k^*$ des relèvements $\omega: J_k U \longrightarrow \Lambda^* W$ de α_k s'identifie, à l'aide de π^* , au module $\Lambda_k^* | J_k U$, l'identification étant compatible avec les différentielles $d_k = d | \Lambda_k^*$. Le lemme ci-dessus est alors une conséquence directe du lemme correspondant énoncé au §2 .

Disons enfin que les calculs locaux du §2 deviennent les calculs locaux sur les Grassmanniennes. En particulier, au k-champ $\xi = \Sigma f_i \, \partial/\partial x^i$ de $J_k U$ correspond le k-champ $\Sigma f_i \, (\partial/\partial x^i + y_i^\lambda \, \partial/\partial y^\lambda)$ de $\mathcal{X}_k$ et l'expression locale (2.1) est également l'expression locale d'un élément de Λ_k^r .

4.- Calcul holonôme dans les variétés Grassmanniennes.

Tout difféomorphisme local $\phi: U \to U'$ de la variété V échange les germes de sous-variétés et par conséquent se prolonge en un difféomorphisme local $p_k \phi : \rho_0^{-1}(U) \longrightarrow \rho_0^{-1}(U')$ de la variété Grassmannienne $G_k V$. De même, tout champ de vecteurs θ défini sur un ouvert U de V se prolonge en un champ $p_k \theta$ de $G_k V$ défini sur $\rho_0^{-1}(U)$. Ces opérations de prolongement sont fonctorielles et préservent le recollement.

Soit ξ un champ de vecteurs et ω une forme différentielle, les deux d'ordre k . On définit le transport et la dérivée de Lie de ξ et ω par θ suivant les formules:

$$(\phi_* \xi)_X = \phi_* (\xi_{p_k \phi^{-1} X}) \qquad L_\theta \xi = \frac{d}{dt} (\phi_{-t})_* \xi \,\Big|_{t=0}$$

$$(\phi^* \omega)_X = \phi^* (\omega_{p_k \phi X}) \qquad L_\theta \omega = \frac{d}{dt} \phi_t^* \omega \,\Big|_{t=0}$$

où (ϕ_t) est une famille locale à un paramètre vérifiant $d/dt \, \phi_t \,|_{t=0} = \theta$. Les dérivées de Lie ci-dessus satisfont aux relations habituelles. On signale en particulier la commutativité $[L_\theta, d] = 0$ ainsi que les *formules fondamentales* (du calcul dans les espaces de prolongement)

$$[L_\theta, \partial_\xi] = \partial_{L_\theta \xi} \qquad [L_\theta, i_\xi] = i_{L_\theta \xi}$$

où i_ξ est le produit intérieur formel et ∂_ξ la dérivée de Lie formelle (cf. [2]) et dont la démonstration est une simple vérification.

5.- Cohomologie des pseudogroupes.

Un champ de vecteurs ξ d'ordre k est invariant par un difféomorphisme local ϕ:

$U \to U'$ lorsque $\phi_* \xi = \xi$; il est invariant par une transformation infinitésimale θ lorsque $(\phi_t)_* \xi = \xi$ où (ϕ_t) est le groupe local à un paramètre engendré par θ . On démontre comme d'habitude, que la condition d'invariance infinitésimale est équivalente à $L_\theta \xi = 0$. De même, une forme différentielle ω d'ordre k est invariante par ϕ (resp. par θ) lorsque $\phi^* \omega = \omega$ (resp. $L_\theta \omega = 0$). Nous dirons, plus généralement, que ξ ou bien ω est invariant par une famille Γ de transformations finies ou infinitésimales locales lorsque l'invariance a lieu par rapport à chaque élément de Γ . Soient $\chi(\Gamma)$ et $\Lambda^*(\Gamma)$ les ensembles de champs et de formes différentielles, de tout ordre, invariants par Γ . Le sous-ensemble $\Lambda^0(\Gamma) = F(\Gamma)$ est le sous-anneau de F dont les éléments sont les invariants différentiels scalaires (de tout ordre) de Γ et $F_k(\Gamma) = F(\Gamma) \cap F_k$ est le sous-anneau des invariants d'ordre k . Les ensembles $\chi(\Gamma)$ et $\Lambda^*(\Gamma)$ sont des modules sur $F(\Gamma)$, $\chi(\Gamma)$ est une sous-algèbre de Lie réelle de χ et, compte tenu des formules du §4 , $\Lambda^*(\Gamma)$ est invariant par les opérateurs d , i_ξ et ∂_ξ avec $\xi \in \chi(\Gamma)$. Par conséquent, le complexe du lemme de Poincaré (cf. §3) se spécialise en le complexe

$$0 \longrightarrow R \longrightarrow \Lambda^0(\Gamma) \xrightarrow{d} \Lambda^1(\Gamma) \xrightarrow{d} \cdots\cdots \xrightarrow{d} \Lambda^p(\Gamma)$$

dont la cohomologie est par définition celle de Γ . Lorsque Γ est un pseudogroupe de Lie dont les groupoïdes associés de tous les ordres $\geq k_0$ sont α-connexes et engendrés par le pseudogroupe infinitésimal $\Theta(\Gamma)$, alors le complexe de cochaînes relatif à la famille Γ est égal à celui relatif à $\Theta(\Gamma)$. L'exactitude locale de ce complexe exige bien sûr certaines conditions de régularité sur la famille Γ et dépend en outre de la structure de Γ comme le montrent les exemples suivants:

Exemple 1. (*Sophus Lie*) Soit $V = R^3$ et Γ le pseudogroupe dont la transformation générale ϕ est donnée par

$$X = \psi(x) \quad , \quad Y = y \quad , \quad Z = z/\psi'(x)$$

où ψ est un difféomorphisme local de R (c'est un exemple de pseudogroupe simple isomorphe au pseudogroupe général de R). Les générateurs infinitésimaux c'est-à-dire, les éléments du pseudogroupe infinitésimal correspondant Θ , sont les champs

$$\theta = f\, \partial/\partial x - f'z\, \partial/\partial z$$

où f est une fonction différentiable arbitraire de la seule variable x et f' est sa dérivée. On pose $p = 2$ et on considère les Grassmanniennes $G_k V$ ainsi que la fibration $\pi: V \to R^2$, $(x,y,z) \longmapsto (x,y)$ et les ouverts affines $J_k V = \bar{G}_k V$. On montre facilement que les champs $\{ \xi_1 = \frac{1}{z} \partial_x , \xi_2 = \partial_y \}$ constituent un repère invariant d'ordre zéro défini dans l'ouvert $J_k V \cap \{z \neq 0\}$ et par conséquent, les formes d'ordre zéro $\{\omega^1 = z\, dx , \omega^2 = dy\}$ constituent un corepère invariant. Chaqu'une de ces formes s'étend globalement en une forme invariante sur V et toute forme invariante est combinaison linéaire, à coefficients dans $F(\Gamma)$, de ω^1 et ω^2 ou bien de $\omega^1 \wedge \omega^2$. On signale en outre que $d\omega^1 = \frac{q}{z}\omega^1 \wedge \omega^2$ et $d\omega^2 = 0$ où $q = \partial z/\partial y$, suivant la notation de Monge. Un calcul très simple montre que le lemme de Poincaré est valable pour ce Γ-complexe comme conséquence du fait que $[\Theta,\Theta] = \Theta$.

Exemple 2. (*Medolaghi*) Soit Γ le pseudogroupe de $V = R^3$ dont la transformation générale s'écrit

$$X = \psi(x) \quad , \quad Y = y\,\psi'(x) \quad , \quad Z = z + y(\psi''(x)/\psi'(x))$$

où ψ est un difféomorphisme local de R . Le pseudogroupe infinitésimal Θ correspondant est l'ensemble des champs

$$\theta = f\,\partial/\partial x + f'y\,\partial/\partial y + f''y\,\partial/\partial z$$

et un système de générateurs pour les formes invariantes est donné par

$$\omega^1 = \frac{1}{y}(dy - z\,dx) \quad , \quad \omega^2 = \frac{1}{y}\,dx$$

défini dans l'ouvert $U = \{y \neq 0\}$ de G_2V . Il n'y a pas de formes invariantes globales non nulles (sauf les fonctions constantes) car les invariants scalaires ne s'annullent pas à l'infini dans la direction de y . Le lemme de Poincaré est valable pour ce Γ-complexe en tout point de U car on a toujours $[\Theta,\Theta] = \Theta$.

Exemple 3. (*Cartan*) Soit Γ le pseudogroupe de $V = R^3$ ensemble des transformations

$$X = \psi(x) \quad , \quad Y = y/\psi'(x) \quad , \quad Z = z/\psi'(x) - \psi''(x)/\psi'(x)^2 \ .$$

Un système de générateurs pour les formes invariantes de G_2V est donné par

$$\omega^1 = y\,dx \quad , \quad \omega^2 = z\,dx - \frac{1}{y}\,dy \ .$$

On démontre, encore dans ce cas, le lemme de Poincaré.

Le même principe nous conduit à la proposition suivante,

Proposition. Soit Θ une famille de champs de vecteurs locaux sur la variété V vérifiant $[\Theta,\Theta] = \Theta$. La cohomologie locale de Θ est alors nulle en dimension un. Cette condition du crochet n'est que suffisante.

6.- Cohomologie des équations différentielles.

Un système d'équations aux dérivées partielles d'ordre k sur la variété V peut être conçu comme étant une sous-variété R_k de G_kV car le point générique de R_k représente un élément de p-sous-variété assujetti à des contraintes infinitésimales d'ordre k . Nous supposerons, pour de raisons techniques, que R_k est une sous-variété régulièrement plongée et que la projection $\rho_0 : R_k \longrightarrow V$ est une fibration (i.e., une surmersion). Une solution de R_k est une sous-variété v de V de dimension p vérifiant $v_k \subset R_k$. En particulier, un système de Pfaff de rang n-p (ou bien une distribution de dimension p) sur la variété V est l'image d'une section globale de $\rho_{01}: G_1V \longrightarrow V$. Le prolongement d'ordre ℓ d'un système R_k est le sous-ensemble $R_{k+\ell} \subset G_{k+\ell}V$ défini par $R_{k+\ell} = G_{k+\ell}V \cap G_\ell R_k$ où $G_\ell R_k$ est la Grassmannienne d'ordre ℓ et de dimension p de la variété R_k et où l'on considère, pour effet de l'intersection, les deux espaces plongés dans $G_\ell G_k V$. La définition montre que $R_{k+\ell} = \{X \in G_{k+\ell}V \mid i_{\ell,k}(X) \in G_\ell R_k\}$. Le prolongé $R_{k+\ell}$ n'est pas en général une sous-variété de $G_{k+\ell}V$. Cependant, lorsqu'il en est ainsi, ce prolongé peut être obtenu par itération de prolongements d'ordre un de même façon que dans le contexte bien connu des variétés de jets. De plus, il est clair que $\rho_{k,k+\ell}R_{k+\ell} \subset R_k$ et que les systèmes prolongés ont les mêmes solutions que R_k . Nous dirons que R_k est *régulier* (par prolongement) lorsque tous les prolongés $R_{k+\ell}$ sont des sous-variétés régulièrement plongées et fibrées sur la base V , et nous supposerons désormais qu'il en soit toujours ainsi.

Indiquons par $\mathcal{X}_{k+\ell}(R_k)$ l'ensemble de toutes les applications différentiables ξ : $R_{k+\ell} \longrightarrow E$ compatibles avec $\rho_{1,k+\ell}$ et par $\Lambda^*_{k+\ell}(R_k)$ l'ensemble des applications différentiables $\omega: R_{k+\ell} \longrightarrow \Lambda E^*$ également compatibles avec $\rho_{1,k+\ell}$. Les éléments de $\mathcal{X}_{k+\ell}(R_k)$ sont appelés les champs de vecteurs d'ordre $k+\ell$ sur R_k et ceux de $\Lambda^*_{k+\ell}(R_k)$ les formes différentielles d'ordre $k+\ell$ sur R_k . Ces espaces sont des modules sur l'anneau des fonctions $F_{k+\ell}(R_k)$ et leures limites inductives $\mathcal{X}(R_k)$ et $\Lambda^*(R_k)$ sont des modules sur $F(R_k) = \varprojlim F_{k+\ell}(R_k)$.

Il est facile de vérifier que le calcul covariant du §3 se restreint aux espaces $\mathcal{X}(R_k)$ et $\Lambda^*(R_k)$, le premier devenant une algèbre de Lie réelle fléchée et le second un module différentiel. On déduit alors un complexe de cochaînes

$$0 \longrightarrow \mathbb{R} \longrightarrow \Lambda^0(R_k) \xrightarrow{d} \Lambda^1(R_k) \xrightarrow{d} \cdots\cdots \xrightarrow{d} \Lambda^p(R_k)$$

dont la cohomologie est par définition celle de l'équation R_k . À toute solution ν de R_k correspond une transformation naturelle (ν_k^*) du complexe précédent en le complexe de *de Rham* de la variété ν et par conséquent une transformation au niveau des cohomologies. Les éléments de l'image de (ν_k^*) sont les *classes fondamentales* associées à la solution ν .

Nous montrons à présent que ces classes se réduisent, dans le contexte Ehresmannien, aux classes fondamentales introduites par Molino dans [5] et [6]. Soit Φ le fibré vectoriel de base G_kV dont la fibre au point X est égale à $TX_{k-1}/i_{1,k-1}X(X_{k-1} = \rho_{k-1,k}X$, $i_{1,k-1}X$ est l'élément de contact holonôme de $G_{k-1}V$ associé à X et TX_{k-1} est l'espace tangent à $G_{k-1}V$ au point X_{k-1}) et $\Omega_k: TG_kV \longrightarrow \Phi$ la *forme fondamentale* de la variété Grassmannienne obtenue par passage au quotient de $T\rho_{k-1,k}$. Dans une carte affine de coordonnées $(x^i,y^\lambda_\alpha)_{|\alpha|\leq k}$, cette forme fondamentale s'écrit

$$\Omega_k = \sum_{0\leq|\alpha|\leq k-1} \omega^\lambda_\alpha \, \partial/\partial y^\lambda_\alpha \qquad \text{où} \qquad \omega^\lambda_\alpha = dy^\lambda_\alpha - y^\lambda_{\alpha+1_i}\, dx^i .$$

On voit ainsi que $\Sigma_k = \mathit{ker}\ \Omega_k$ n'est autre que le système de contact canonique sur

G_kV qui à chaque point X associe le sous-espace vectoriel engendré par les éléments de contact linéaires tangents aux sections holonômes (i.e., de la forme $y \in \nu \longmapsto \nu_k y$) qui passent par X . Étant donné un système R_k , on indique par $\Omega(R_k)$ la restriction de Ω_k à la sous-variété R_k et par $\Sigma(R_k)$ le champ d'éléments de contact linéaires induit sur R_k par Σ_k, autrement dit, le champ $\mathit{ker}\ \Omega(R_k)$. Faisons maintenant les hypothèses suivantes:

a) Le prolongement R_{k+1} est une sous-variété et $\rho_{k,k+1}: R_{k+1} \longrightarrow R_k$ est surjectif (i.e., R_k est 1-intégrable),

b) Le champ $\Sigma(R_k)$ est de dimension constante.

Reprenons maintenant les définitions de Molino (cf. [6]) et restreignons les considérations précédentes à l'Ehresmannienne J_kV où $\pi: V \to W$ est une fibration donnée, $\dim W = p$ et $J_kV = \bar{G}_kV$ est un ouvert dense de G_kV . Soit A l'algèbre des formes différentielles extérieures de la variété R_k et I l'idéal *caractéristique* de $\Sigma(R_k)$ c'est-à-dire, l'idéal engendré par les formes différentielles linéaires nulles sur $\Sigma(R_k)$ ainsi que par leurs différentielles extérieures (on utilise ici le calcul différentiel extérieur habituel sur la variété R_k). On définit à l'aide du relèvement holonôme (et en se plaçant dans le contexte semi-basique du §2) un morphisme linéaire injectif *de retour à la base*

$$H: A/I \longrightarrow \Lambda^*_{k+1}(R_k)$$

$H[\omega] = \omega \circ \Lambda H_{k+1}$ où $[\omega]$ est la classe de ω . Il est clair que toute forme linéaire ω de I s'annule par H et, puisque le système dérivé de Σ_{k+1} est égal à $(\rho_{k,k+1})^{-1}_*\Sigma_k$, la même conclusion suit pour $d\omega$. De plus, si $H\omega = H\mu$ alors $H(\omega-\mu) = 0$ c'est-à-dire, $\omega-\mu$ est dans l'idéal engendré par les formes caractéristiques linéaires. Montrons enfin que $H\, d[\omega] = d\, H[\omega]$. Il suffit de faire un calcul local qui de fait est le plus suggestif, l'argument global faisant appel au système dérivé de Σ_{k+1} . Or, si $\omega = f$ est une fonction, alors $H f = f$,

$$df = \partial f/\partial x^i\, dx^i + \cdots\cdots + \partial f/\partial y^\lambda_\alpha\, dy^\lambda_\alpha + \cdots \quad , \quad |\alpha| \leq k \ ,$$

et

$$H\, df = \partial f/\partial x^i\, dx^i + \cdots\cdots + \partial f/\partial y^\lambda_\alpha\, y^\lambda_{\alpha+1_i}\, dx^i + \cdots = \Sigma\, \partial_i f\, dx^i = df \ .$$

Si ω est une forme linéaire, alors

$$\omega = f_i\, dx^i + \cdots\cdots + f^\alpha_\lambda\, dy^\lambda_\alpha + \cdots \quad , \quad |\alpha| \leq k \ ,$$

$$H\omega = f_i\, dx^i + \cdots\cdots + f^\alpha_\lambda\, y^\lambda_{\alpha+1_i}\, dx^i + \cdots \ ,$$

$$d\omega = df_i \wedge dx^i + \cdots\cdots + df^\alpha_\lambda \wedge dy^\lambda_\alpha + \cdots \ ,$$

$$H\, d\omega = df_i \wedge dx^i + \cdots\cdots + df^\alpha_\lambda \wedge y^\lambda_{\alpha+1_i}\, dx^i + \cdots \ ,$$

$$d\, H\omega = df_i \wedge dx^i + \cdots\cdots + df^\alpha_\lambda \wedge y^\lambda_{\alpha+1_i}\, dx^i + f^\alpha_\lambda\, dy^\lambda_{\alpha+1_i} \wedge dx^i + \cdots$$

d'où l'égalité $H\, d\omega = d\, H\omega$ car

$$dy^{\lambda}_{\alpha+1_i} \wedge dx^i = y^{\lambda}_{\alpha+1_i+1_j} dx^j \wedge dx^i = 0 .$$

La conclusion s'étend aux formes de degré r car H commute avec le produit extérieur et les opérateurs d et d sont des dérivations de degré un. On voit donc, H étant injectif et commutant avec les deux différentielles, que les *cocycles* ainsi que les *cobords* se correspondent par H . Ceci achève la démonstration. En particulier, lorsque nous considérons un système R_1 définissant un feuilletage (système intégrable défini par l'image d'une section $\sigma: V \to G_1 V$), on retrouve les classes de Bott-Haefliger comme le fait voir Molino à l'aide de sa construction des classes fondamentales ([6]). De même, on retrouve les classes définies dans [3],[4],[8] et, plus généralement, on peut appliquer ces méthodes aux Γ-structures. Cependant, l'aspect qui semble le plus intéressant est celui où l'on se donne un système R_k invariant par l'action d'un pseudogroupe fini ou infinitésimal Γ (système de Lie-Vessiot) auquel on appliquera simultanément les méthodes des deux derniers paragraphes.

7.- Cohomologie équivariante.

Soit R_k un système différentiel régulier et invariant par une famille Γ de transformations finies ou infinitésimales c'est-à-dire, la sous-variété R_k est invariante par les prolongés des éléments de Γ . On montre facilement que les prolongés $R_{k+\ell}$ sont aussi invariants, ce qui permet de ne considérer sur R_k que les formes différentielles d'ordre $k+\ell$ qui sont invariantes par Γ . La cohomologie du complexe ainsi déterminé est par définition la cohomologie équivariante de l'équation R_k et les classes fondamentales associées aux solutions ν sont les classes équivariantes.

Pour calculer effectivement les cohomologies définies précédemment, il conviendra connaître des méthodes permettant la détermination de formes différentielles Γ-invariantes. De telles méthodes se trouvent par exemple dans [1], chap.XII, et dans [2].

BIBLIOGRAPHIE

[1] E. Cartan. La théorie des groupes finis et continus et la géométrie différentielle traitées par la méthode du repère mobile. Gauthier-Villars. Paris. 1951.

[2] A. Kumpera. Invariants différentiels d'un pseudogroupe de Lie, I-II. J. Differ. Geom. 10,(1975),279-335,347-415.

[3] P. Libermann. Sur les systèmes de Pfaff totalement réguliers. Journées Franco-Espagnoles. Toulouse. (1975).

[4] J. Martinet. Classes caractéristiques des systèmes de Pfaff. Lecture Notes in Math. 392. Springer-Verlag. Berlin, Heidelberg. (1974),30-37.

[5] P. Molino. Γ_q-structures partielles et classes de Bott-Haefliger. Note C.R.Acad. Sc. Paris. 281,(1975),203-206.

[6] P. Molino. Cohomologie partielle et équations différentielles. Géométrie Différentielle, Université Paris VII. Paris. (1978),111-119.

[7] J. Pradines. Théorie de Lie pour les groupoïdes différentiables. Calcul différentiel dans la catégorie des groupoïdes infinitésimaux. Note C.R. Acad. Sc.

Paris. 264,(1967),245-248.

[8] J. Pradines. Remarque sur le théorème d'annulation de Bott-Martinet. Note C.R. Acad. Sc. Paris. 282,(1976),527-529.

[9] W. Tulczyjew. The Euler-Lagrange resolution. International Colloq. Differ. Geom. Methods in Math. Physics. Aix-en-Provence. (1979).

[10] A. Vinogradov. Geometry of non-linear differential equations. J. Soviet Math. 17,(1981),1624-1649.

ENERGIES ET GEOMETRIE INTEGRALE

R. Langevin
Départament de Mathématiques
Faculté des Sciencies - Mirande
Université de Dijon
21004 Dijon. France

Plutôt que de répéter ici des résultats déjà écrits, je souhaiterais insister sur quelques difficultés que l'on rencontre en cherchant à associer une énergie à un feuilletage $\mathcal{F}$ d'une variété riemannienne, puis sur d'autres difficultés que l'on rencontre en essayant d'étudier $\mathcal{F}$ à l'aide de coupes.

1.- Energie d'un feuilletage.

Soit $\mathcal{F}$ un feuilletage d'une variété M compacte munie de la métrique g.

Le feuilletage $\mathcal{F}$, et même plus généralement un champ de p-plans $\mathcal{P}$, définit une projection du fibré tangent à M, TM sur le fibré normal à $\mathcal{F}$: $Q = TM/T\mathcal{F}$ ($T\mathcal{F}$ est le sous-fibré de TM dont la fibre $T_x\mathcal{F}$ est l'espace tangent en x à la feuille de $\mathcal{F}$ passant par x).

On peut définir l'énergie $E(\mathcal{F})$ de $\mathcal{F}$ comme étant l'énergie de cette projection (voir [Ka-To]$_1$, [Ka-To]$_3$ et [Ee-Sa]$_1$ pour une définition de l'énergie d'une application, ainsi que pour une étude des applications harmoniques, c'est-à-dire points critiques de l'énergie).

Cependant un feuilletage est souvent défini à l'aide d'une famille de cartes U_i chacune munie d'une submersion $p_i : U_i \longrightarrow \mathcal{T}_i$. Nous pouvons, lorsque $\mathcal{T}_i$ est muni d'une métrique calculer, (voir [Ee-Sa]$_1$), l'énegie de la submersion p_i.

Il serait souhaitable de pouvoir calculer l'énergie du feuilletage $\mathcal{F}$ à l'aide d'une famille quelconque de cartes.

C'est possible lorsque le feuilletage $\mathcal{F}$ est riemannien; c'est-à-dire, admet une métrique transverse invariante.

Proposition 1 (implicite chez [Ka-To]$_{1,2,3}$).- Munissons chacune des transversales $\mathcal{T}_i$ de la métrique transverse invariante, et soit ψ_i une partition de l'unité subordinnée au recouvrement de M par les ouverts

U_i, on a:

$$E(\mathcal{F}) = \sum \psi_i \, E(p_i).$$

Remarque 1.- Si l'on réalise les transversales $\mathcal{T}_i$ par des sous-variétés contenues dans U_i que l'on muni de la métrique induite, le résultat est à corriger de la différence de l'énergie des plongements ($\mathcal{T}_i$, métrique invariante) $\longrightarrow$ M et du volume de $\mathcal{T}_i$ muni de la métrique invariante.

2.- Etudier la somme des énergies associés à deux feuilletages ou champ de plans orthogonaux devrait être analogue à l'étude de l'énergie de F section de fibrés faite dans [Ee-Sa]$_2$.

Problème 1,(peut-être mal posé).- Exprimer à l'aide des cartes de l'énergie d'un feuilletage quelconque.

Problème 2.- Expliciter $E(\mathcal{F})$, ou $E(\mathcal{F}) + E(\mathcal{F}^{\perp})$ à l'aide de secondes formes fondamentales de $\mathcal{F}$ et $\mathcal{F}^{\perp}$. Peut-être si $\mathcal{F}^{\perp}$ n'est pas intégrable faut-il considerer une seconde forme fondamentale non symétrique,(voir [Re]$_2$ et [La]$_3$).

2.- Energie d'un cristal liquide. (voir [De Ge])

Un cristal est un arrangement de molécules dont les centres occupent, en première approximation, des positions déterminées par un reseau à 3 dimensions. Tandis que dans un liquide ordinaire les molécules sont "en désordre", dans un cristal liquide celles-ci sont ordonnées suivant les feuilles d'un feuilletage de dimension 1(nematiques) ou 2 (smectiques)
Qui plus est, lorsque l'on schématise les molécules d'un cristal liquide par un batonnet, disons le plus grand diamètre de l'ellipsoïde d'inertie de celle-ci, les directions définies par ces batonnets varient lentement, (disons que l'angle formé par deux batonnets voisins est petit par rapport à leur distance). En d'autres termes, à un cristal liquide, est associé un champ de droites $\mathcal{D}$ sur la région W de $\mathbb{R}^3$ remplie par le cristal liquide.

Lorsque le cristal liquide est un un nématique ce champ est tangent au feuilletage de dimension 1 qui ordonne les molécules, (ou sert-il a le définir ?).

Les différents smectiques sont classés suivant la position de ce champ par rapport au feuilletage de dimension 2 qui définit l'ordre.

Remarque.- Comme les molécules d'un smectique sont réparties en couches d'épaisseur constante il est intéressant à ce sujet d'étudier les feuilletages mesurés. (voir [Po]).

Imaginons qu'un nématique "idéal" soit ordonné suivant un feuilletage $\mathcal{F}$ formé de droites parallèles. Si nous lui imposons des contraintes, (champ magnétique, échauffement,...), ce feuilletage $\mathcal{F}$ est déformé.

Les physiciens associent à cette situation une énergie libre F_d. Localement au moins le champ $\mathcal{D}$ peut être décrit par un champ de vecteurs unitaire $N(x)$. Posons

$$F_d(x) = C_1(\operatorname{div} N)^2 + C_2(N\cdot \operatorname{rot} N)^2 + C_3 \|N \wedge \operatorname{rot} N\|^2$$

où C_1, C_2, C_3 sont des constantes physiques souvent d'ordre de grandeur comparables.

Dans $[La]_3$ j'interprète, ne faisant là que recopier une partie d'un travail de Rogers [Ro] les trois termes de cette formule géométriquement. $\|N \wedge \operatorname{rot} N\|^2$ correspond au carré de la courbure des feuilles de $\mathcal{F}$, $(N.\operatorname{rot} N)^2$ à la non intégrabilité du champ de plans $\mathcal{P}$ orthogonal à $\mathcal{F}$, $(\operatorname{div} N)^2$ au carré de la courbure moyenne de $\mathcal{P}$.

Il reste à calculer explicitement le lien entre ces quantités et l'énergie du feuilletage $\mathcal{F}$ et du champ de plans $\mathcal{P}$.

Cette étude a été suscité par une note de Cladis Kléman et Pieranski qui montrent par une "décoration technique" un feuilletage qui s'avère être presque le revêtement d'un des feuilletages défini par un champ de repères harmonique sur le tore T^2, (voir $[La]_3$ et $[La]_2$ pour une autre figure ressemblant à celle de [Cla-Kle-Pi])

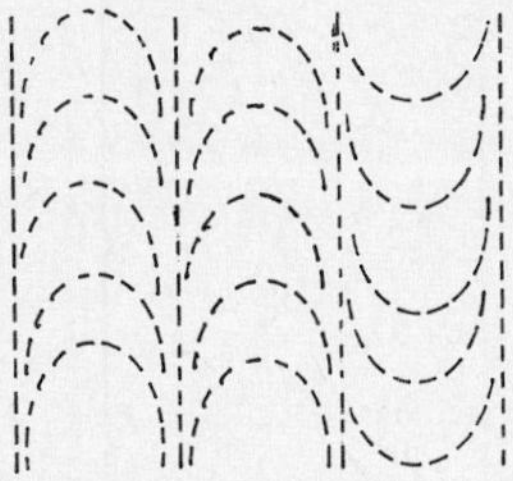

"décoration technique"
de [Cla-Kle-Pi]

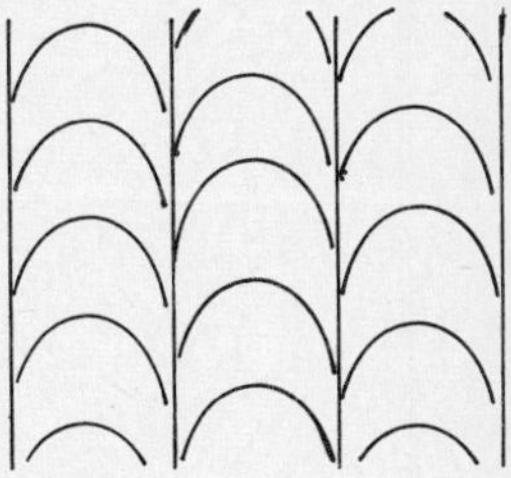

Feuilletage relèvement à $\mathbb{R}^2$ d'un feuilletage défini par un champ de repères harmonique sur T^2.

Problème.- Lier ces "énergies" aux nombres $E(F)$ et $E(F^{\perp})$ définis plus haut.

3.- Energies et points singuliers.

Soit F un feuilletage d'une surface. Notons k la courbure géodésique des feuilles de F.

Un défaut de l'intégrale de courbure $\int k^2$ est que celle-ci devient infinie dès que le feuilletage F admet de singularités.

Donnons un exemple pour préciser ce dernier point.

Soit F un feuilletage d'une surface S admettant des singularités isolés telles que le feuilletage au voisinage d'un point singulier soit diffeomorphe au feuilletage défini par les niveaux de $\mathrm{Re}\, z^k = c$ où $|z| = c$, $z \in \mathbb{C}$.

On a $$\int_{M-\Sigma} k^2 = \infty$$

(Σ est l'ensemble des points singuliers de F)

Il est toutefois encore possible de définir des feuilletages harmoniques avec singularités en se débarassant de la "partie infinie" de l'intégrale.

Remarquons d'abord que la contribution locale de la singularité $\{\mathrm{Im}\, z^k = c\}$ à l'integrale de la somme des carrés des courbures géodésiques de F et $F^{\perp}$ est

$$\int_{B(0,r_1)-B(0,r_2)} k_1^2 + k_2^2 = 2\pi |\text{indice } 0| \cdot \mathrm{Log}\, \frac{r_1}{r_2} \qquad (r_1 > r_2)$$

Il est donc naturel de dire qu'un feuilletage d'une surface dont les singularités sont données avec leur indice est harmonique si la limite

$$I = \lim_{r \to 0} \int_{S-\Sigma_r} k_1^2 + k_2^2 + 2\pi \sum_{x \in \Sigma} |\text{indice } (x)|\ \mathrm{Log}\, r$$

est minimale. ($\Sigma_r = \bigcup_{x \in \Sigma} B(x,r)$).

Remarque.- Cette limite n'a pas de sens en elle-même puisque le choix de la "renormalisation" est fait à une constante arbitraire près.

Problème 1.- Y-a-t-il une manière naturelle de "renormaliser" cette intégrale?

Problème 2.- Pourrait-on tirer une information de l'intégrale $\int_S \sqrt{k_1^2 + k_2^2}$ qui elle converge.

Remarque.- Les défauts d'un cristal liquide qui correspondent aux singularités du feuilletage ont aussi été étudiés d'un point de vue topologique, (voir par exemple [Po] et [Me]).

4.-Energies et géométrie intégrale

Ce paragraphe est le récit de l'échec d'une tentative d'étudier à l'aide de la geométrie integrale les facteurs de l'énergie libre d'un cristal liquide nématique defini au §2.

Donnons d'abord une propriété des surfaces plongées dans l'espace euclidien.

Soit S une surface éventuellement à bord, compacte plongée dans $\mathbb{R}^3$.

Notons k(x) la courbure en x de la courbe du plan H intersection de S avec le plan affine H.

Notons $\sigma_1(x)$ la courbure moyenne de S en x et K(x) la courbure de Gauss de S en x.

Proposition (Je ne connais pas la réference)

$$\int_{\mathcal{A}}\int_{H\cap S} (k(x))^2 = A\int_S (\sigma_1(x))^2 + B\int_S K(x)$$

où A et B sont des constantes ne dépendant que des dimensions et $\mathcal{A}$ l'ensemble des plans affines de $\mathbb{R}^3$ muni de sa mesure canonique.

Malheureusement, s'il est facile de couper un champ de plans P défini sur un ouvert W de $\mathbb{R}^3$ par un plan affine, le résultat de la proposition n'est plus vrai car pour un ensemble de mesure non nulle de $\mathcal{A}$ le feuilletage $(P\cap H)$ admet des singularités.

Probleme.- Comment éliminer une "contribution divergente" de façon à obtenir une relation entre

$$\int_W \sigma_1^2(x) \ , \ \int_W K(x) \ , \ \int_W (\text{terme de non integrabilité})^2$$

et une intégrale remplaçant $\int_A \int_{W \cap H} (k(x))^2$?

Le cas des cristaux liquides cholestériques est encore plus particulier.

a) Les molécules sont disposées en couches.

b) Le champ de droites défini par celles-ci est tangent à ces couches.

La configuration d'énergie libre minimale d'un cholestérique est donnée par le champ

$$N = \begin{pmatrix} \cos az \\ \sin az \\ 0 \end{pmatrix}$$

Le champ N est constant dans les plans horizontaux et tourne à vitesse constante lorsque l'on suit un axe vertical.

Le champ $\mathcal{P} = (N)^{\perp}$ découpe sur tout plan affine ne contenant pas l'axe vertical un feuilletage "en arceaux".

Pour $\theta =$ angle (Oz,H) petit ces sections sont de la forme

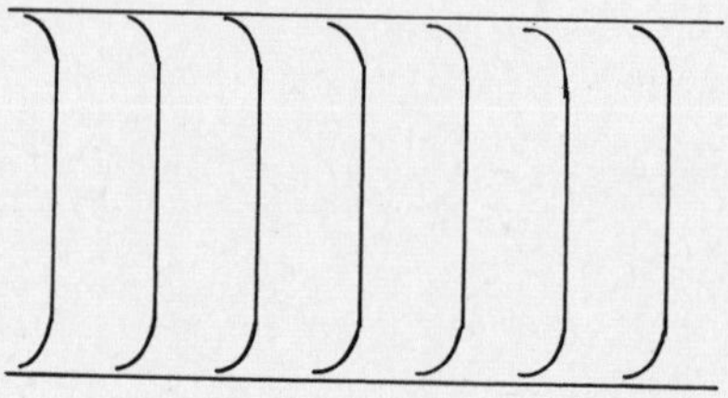

En effet lorsque l'angle θ tend vers 0 le feuilletage $\mathcal{P} \cap H$ tend vers la configuration :

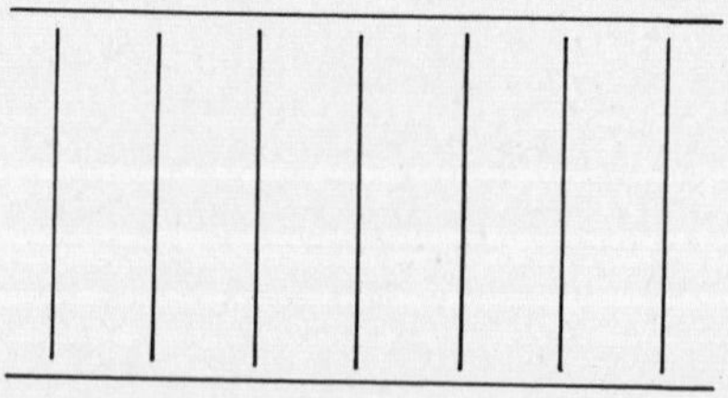

L'ensemble des plans affines contenant un axe vertical est de mesure nulle mais lorsque l'angle θ est petit la courbure des feuilles de $P \cap H$ est de l'ordre de $1/\theta$ sur deux bandes d'epaisseur θ voisine de chaque feuille horizontale de $P \cap H$, ce qui implique que l'integrale $\int_H (k(x))^2$ est de l'ordre de $\frac{1}{\theta}$ tandis que la mesure de l'ensemble des plans θ proche de la vertical est elle aussi de l'ordre de θ ce qui conduit pour estimer $\int_A \int_{W \cap H} (k(x))^2$ à calculer une intégrale de l'ordre $\int_0^{\varepsilon} \frac{d\theta}{\theta}$.

Problème.- Peut-on dans ce cas éliminer la contribution des plans affines presque verticaux à l'intégrale $\int_A \int_{H \cap W} (k(x))^2$?

En résumé si l'on étudie un champ de plans P défini sur un ouvert de $\mathbb{R}^3$ l'intégrale $\int_A \int_{H \cap W} (k(x))^2$ semble diverger en général.

Définissons l'application de Gauss γ_P associée au champ P. L'application γ_P est à valeurs dans le fibré en droites canonique sur $\mathbb{P}$: $E = \{D \in \mathbb{P}_2 , u \in D \}$. Posons

$$\gamma_P(x) = \text{(droite D normale en x à } P\text{, projection orthogonale u de x sur D)}$$

Le fibré en droites E est isomorphe à l'espace des plans affines de $\mathbb{R}^3$ par l'application $H \longrightarrow (D = H^{\perp} , u = D \cap H)$

Suivant la mesure de $\gamma_P(W)$ l'intégrale $\int_{A \; W \cap H} (k(x))^2$ diverge pour une raison différente.

Cas 1.- Mesure $\gamma_P(x) \neq 0$. On a alors un ensemble de mesure non nulle de plans affines sur lesquels la trace de P admet des ingularités.

Cas 2.- Mesure $\gamma_P(W) = 0$.

a) La seconde forme fondamentale de P est de trace non nulle sur W. On peut montrer que le champ P est constant le long d'un feuilletage en droites N et est tangent à N. L'ensemble $\gamma_P(W)$ est de dimension 2. La preuve est analogue à celle de l'existence de génératrices.

sur une surface de courbure de Gauss nulle, voir [DoCa].

b) La seconde forme fondamental de P est nilpotente. Ce cas a été considéré par Bouligand et Chenciner , (Bouligand à paraitre).

Mais l'étude du champ $\begin{pmatrix} \cos az \\ \sin az \\ 0 \end{pmatrix}$ montre qu'il n'y a aucune raison pour que l'intégrale $\int_A \int_{W \cap H} (k(x))^2$ converge en général.

Il reste donc à espérer qu'une "renormalisation" convenable permette de comprendre le lien entre les énergies que l'on peut associer à P et celles de ses sections.

BIBLIOGRAPHIE

[A] D. Asimov, Average Gaussian curvature of leaves of foliations, Bull. of the Amer. Soc. Vol. 48, (1978), 131-133.

[B.L.R.] F. Brito, R. Langevin, H. Rosenberg, Intégrales de courbure sur une variétée feuillétée, note C.R. Acad. Sc., série A, 285, (1977), 533-536.

[C.K.P.] P. Cladis, M. Kleman, P. Pieranski, note C.R. Acad. Sc., série B, 273, (1971), 275-277.

[DoCa] M. Do Carmo, Differential geometry of curves and surfaces. Prentice Hall. (1976).

$[\text{Ee-Sa}]_1$ Eells & Sampson, Harmonic mappings of riemannian manifolds, Amer. Journal of Math. 86, (1964), 109-160.

$[\text{Ee-Sa}]_2$ Eells & Sampson, Varational theory in Fibre bundles, Proc, US-Japan, seminar of Differential geometry, Kyoto (1965), 61-69.

[De Ge] P.G. de Gennes, liquid cristals, Clarendon Press, Oxford university press, 1974.

$[\text{Ka-To}]_1$ F.W. Kamber & P. Tondeur, Feuilletages harmoniques, note C.R. Acad. Sc. Paris, série A, 291, 409-411.

$[\text{Ka-To}]_2$ F.W. Kamber & P. Tondeur, Infinitesimal automorphisms and second variation of the energy for harmonic foliations, preprint. (1982).

[Ka-To] F.W. Kamber & P.Tondeur. Curvature properties of harmonic foliations, preprint (1982).

$[\text{La}]_1$ R. Langevin, Thèse, publications mathématiques d'Orsay, 431

$[\text{La}]_2$ R. Langevin, feuilletages tendus, Bulletin de la sociétée mathématique de France. 107, (1979), 271-281.

[La]$_3$ R. Langevin, Feuilletages, énergies et cristaux liquides, Conférence Mulhouse 1982,(à paraitre).

[La-Le] R. Langevin & G. Levitt, Courbure totale des feuilletages des surfaces, Commentarii mathematici Helvetici. 57,(1982), 175-195.

[L-S] R. Langevin & T. Shifrin, Polar varities, American Journal of Mathematics. 104, (1982),533-605.

[Me] N.D. Mermin, The topological theory of defects in ordered media, Review of modern physics. 51, (1979), 591-648.

[Po] V. Poenaru , Some aspects of the theory of defects of ordered media and gauge fields related to foliations, Communications in mathematical physics. 80, (1981), 127-136.

[Re] B.L. Reinhart, The second fundamental form of a plane field, Journal of differential geometry, 12, (1977), 619-627.

[Re-Wo] B.L. Reinhart & J.W. Wood, A metric formule for the Gobdillon-Vey invariant for foliations, Proc. of the Amer. Math. Soc. 38, (1973), 427-430.

[Ro] R.P. Rogers, Some differential properties of the orthogonal trajectories of a convergence of curves, with an application to curl and divergence of vectors, Proceeding of the Royal Irish Academy. Sect. A, 29, N°6, (1932).

GEOMETRY AND COHOMOLOGIES ASSOCIATED WITH A CONTACT MANIFOLD

A. Lichnerowicz
Collège de France
Paris

In this lecture, my purpose is to study by geometrical means, the cohomology corresponding to the Lie derivative of the Lie algebra of the infinitesimal automorphisms of a contact manifold, with values in the scalar densities. We prove the existence of an important cohomology 2-class β_1, always $\neq 0$, which is a universal invariant of the contact structures. This invariant plays an important role in the theory of the pseudo differential operators of order one, with real symbols, on an arbitrary manifold. For a compact manifold, such a theory has been given by Omori and coll. [4] (see also [12]).

1 - Pfaffian manifolds and compact manifolds.

a) Let $\hat{W}$ be an oriented smooth differentiable manifold of dimension $m = 2n + 1$. All the introduced elements are supposed of C^∞-class. We set in particular $\hat{N} = C^\infty(\hat{W};\mathbb{R})$. The manifold $\hat{W}$ admits an m-form density of weight -1, denoted by $\hat{T}$, such that, in the domain $\hat{U}$ of a positive chart $\{x^a\}$ $(a,b,\ldots = \bar{0},1,\ldots,2n)$, we have

$$\hat{T}_{|\hat{U}} = dx^{\bar{o}} \wedge dx^1 \wedge \ldots \wedge dx^{2n}$$

Let $\{x^{b'}\}$ be another positive chart of domain $\hat{V}$. If $\hat{U} \cap \hat{V} \neq \emptyset$, we set $A^a_{b'} = \partial x^a / \partial x^{b'}$ and we denote by $J = J^{\hat{U}}_{\hat{V}} = \det(A^a_{b'})$ the jacobian associated with the two charts. We have : $\hat{T}_{\hat{V}} = J^{-1}\hat{T}_{\hat{U}}$. A scalar density $\hat{k}$ of weight 1 -or kernel- is defined on each positive chart of an atlas by one component, these components satisfying

$$\hat{k}_{\hat{V}} = J\,\hat{k}_{\hat{U}} \tag{1-1}$$

It follows that $\hat{k}\hat{T}$ is an usual m-form. A kernel $\hat{k}$ is said to be positive if its components are strictly positive. We consider here only positive kernels and positive m-forms.

b) A pfaffian structure is defined on $\hat{W}$ by a 1-form $\hat{\omega}$ such that $\hat{\eta} = \hat{\omega} \wedge (d\hat{\omega})^n \neq 0$ is a positive m-form; $\hat{\eta}$ determines by means of $\hat{\eta} = \hat{k}\hat{T}$ a positive kernel. The pfaffian manifold $(\hat{W}, \hat{\omega})$ admits a fundamental vector field $\hat{E}$ (or Reeb vector field) given by the relations (where i(.) is the inner product):

$$i(\hat{E})\,\hat{\omega} = 1 \qquad i(\hat{E})\,\hat{F} = 0 \qquad (\text{with } \hat{F} = d\hat{\omega}) \tag{1-2}$$

and an antisymmetric contravariant 2-tensor $\hat{\Lambda}$ given by:

$$\hat{\Lambda}^{ab}\,\hat{\omega}_a = 0 \qquad \hat{\Lambda}^{ab}\,\hat{F}_{ac} = \delta^b_c - \hat{E}^b\,\hat{\omega}_c \tag{1-3}$$

The pair $(\hat{E}, \hat{\Lambda})$ which is such that $\hat{E} \wedge \hat{\Lambda}^n \neq 0$ everywhere, satisfies the relations [1]:

$$[\hat{E}\,,\,\hat{\Lambda}] = 0 \qquad [\hat{\Lambda}\,,\,\hat{\Lambda}] = 2\,\hat{E} \wedge \hat{\Lambda} \tag{1-4}$$

where [,] is the Schouten bracket. Conversely such a pair defines a pfaffian structure.

I have proved that there exist on $(\hat{W}, \hat{\omega})$ linear connections $\hat{\Gamma}$ without torsion such that we have, for the corresponding covariant differentiation $\hat{\nabla}$:

$$\hat{\nabla}\hat{E} = 0 \qquad \hat{\nabla}\hat{F} = 0$$

Such a connection is called a pfaffian connection [3].

c) Two pfaffian structures $\hat{\omega}$ and $\hat{\omega}'$ are said to be equivalent if there is $\phi \in \hat{N}$ such that :

$$\hat{\omega} = e^{\phi}\,\hat{\omega}' \tag{1-5}$$

An (oriented) contact structure is defined on $\hat{W}$ by an equivalence class of pfaffian structures. If $\hat{\omega}$ and $\hat{\omega}'$ are two equivalent pfaffian structures, we have for the corresponding kernels $\hat{k}$ and $\hat{k}'$:

$$e^{\phi} = \hat{k}^{1/(n+1)}\hat{k}'^{-1/(n+1)} \qquad \phi = \frac{1}{n+1}(\log \hat{k} - \log \hat{k}') \quad (1\text{-}6)$$

It follows from (1-5) and (1-6) :

$$\hat{k}'^{-1/(n+1)}\,\hat{\omega}' = \hat{k}^{-1/(n+1)}\,\hat{\omega}$$

We are led to introduce the 1-form density of weight $-1/(n+1)$, given by $\hat{\pi} = \hat{k}^{-1/(n+1)}\hat{\omega}$; $\hat{\pi}$ satisfies the relation :

$$\hat{\pi} \wedge (d\hat{\pi})^n = \hat{T} \qquad (1\text{-}7)$$

Conversely such a 1-form density defines a contact structure. We have :

<u>Proposition.- A contact structure is defined on $\hat{W}$ by a 1- form density $\hat{\pi}$, of weight $-1/(n+1)$, satisfying the relation (1-7).</u>

Here we define thus the contact structures, in general without reference to corresponding pfaffian structures. <u>An infinitesimal automorphism</u> (i. a) of the contact manifold $(\hat{W},\hat{\pi})$ is given by a vector field $\hat{X}$ of $\hat{W}$ such that $L(\hat{X})\hat{\pi} = 0$, where $L(\ .\ )$ is the Lie derivative. We denote by $\hat{L}$ the Lie algebra af the i. a of the contact structure. If $(\hat{E},\hat{\Lambda})$ defines a pfaffian structure, the elements of the corresponding Lie algebra $\hat{L}$ are given by :

$$\hat{X}_{\hat{u}} = \hat{u}\,\hat{E} + [\hat{\Lambda}, \hat{u}] \qquad (\hat{u} \in \hat{N}) \qquad (1\text{-}8)$$

<u>2.- Contact manifold and corresponding exact sympletic manifold.</u>

a) Let $(\hat{W}, \hat{\pi})$ be a contact manifold. Introduce the canonical bundle $p : W \to \hat{W}$ of the <u>positive</u> m-forms of $\hat{W}$. This bundle, of dimension $m+1 = 2n+2$, admits a canonical m-form λ defined at each point y of W such that $py = \hat{x}$ by :

$$\lambda_y\,(X_1,\ldots,X_m) = y(p_*X_1,\ldots,p_*X_m) \qquad (2\text{-}1)$$

where $X_1,\ldots,X_m \in T_yW$ and so $p_*X_1,\ldots,p_*X_m \in T_{\hat{x}}\hat{W}$. If $\{x^a\}$ is a chart of $\hat{W}$ of domain $\hat{U}$, it defines a chart $\{y^A\} = \{y^\circ > 0,\ y^a = x^a\}$ $(A,B = 0,\overline{0},1,\ldots,2n)$ of W of domain $p^{-1}(\hat{U})$ such that :

$$\lambda|_{p^{-1}(\hat{U})} = y^{\circ}\, dy^{\bar{o}} \wedge dy^{1} \wedge \ldots \wedge dy^{n} \tag{2-2}$$

Substitute to y° the coordinate z° given by :

$$y^{\circ} = e^{(n+1)z^{\circ}} \qquad (z^{\circ} \in \mathbb{R})$$

$e^{z^{\circ}}$ has the variance of a scalar density of weight $1/(n+1)$ and determines thus on W such a density denoted by the same notation. Consider on W the 1-form given by :

$$\omega = e^{z^{\circ}}\, p^{*}\hat{\pi} \tag{2-3}$$

which is canonical for the contact structure. It is clear that (W,ω) is an exact symplectic manifold : if $F = d\omega$, we have $F^{n+1} = d\lambda$. We obtain :

<u>Theorem - A contact manifold $(\hat{W}, \hat{\pi})$ being given, its canonical bundle of the positive $(2n + 1)$-forms $p : W \to \hat{W}$ admits a canonical structure ω of exact symplectic manifold, for which ω is $\neq 0$ everywhere</u>.

Arnold has given another "symplectization" of a contact manifold which is, of course, isomorphic to the former one. It is simpler, for various geometrical and cohomological problems, to work at the same time on the contact manifold and on the exact symplectic manifold. We set $N = C^{\infty}(W;R)$. The fundamental vector field Z of (W,ω) is determined by :

$$i(Z)F = \omega$$

and admits in a chart $\{z^{\circ}, z^{a} = x^{a}\}$ only the non vanishing component $Z^{\circ} = 1$; Z is thus the field of the homotheties of the vector bundle $p : W \to \hat{W}$. We have $i(Z)\omega = 0$ and so :

$$L(Z)\,\omega = \omega \tag{2-4}$$

To a pfaffian connection on $\hat{W}$ corresponds on (W,ω) <u>a symplectic connection Γ invariant under Z</u> (see [3]).

b) We denote by $\mu : TW \to T^{*} W$ the isomorphism of vector bundles defined by $\mu(V) = - i(V)F$; μ can be extended to tensors in a natural way and $\Lambda = \mu^{-1}(F)$ is the structure tensor of the symplectic manifold (W,F) ;Λ satisfies :

$$L(Z)\,\Lambda = -\Lambda \tag{2-5}$$

if $u \in N$, we denote by X_u the hamiltonian vector field $\mu^{-1}(du) = [\Lambda, u]$ determined by $u \in N$.

Let L_ω be the Lie algebra of the vector fields of W which preserve ω . Such a field is invariant under Z and so is projectable by p which defines a canonical isomorphism between the Lie algebras L_ω and $\hat{L}$. The elements of L_ω are the hamiltonian vector fields corresponding to the solutions $u \in N$ of the equation :

$$L(Z)\; u = u \tag{2-6}$$

Let N_1 be the space of these solutions. The Poisson bracket $\{\ ,\ \}$ of (W,F) given by Λ defines on N_1 a Lie algebra structure and the algebra $(N_1, \{\ ,\ \})$ is isomorphic to L_ω and so to $\hat{L}$. We consider always N_1 with its Lie algebra structure.

3.- The space N_h and the cohomologies with values in N_h.

a) We denote by N_h (where $h \in \mathbb{R}$) , the space of solutions of the equation :

$$L(Z)\; u = h\, u \tag{3-1}$$

We say that an element of N_h is homogeneous of degree h. It follows from (2-5) that the Lie algebra $(N_1, \{\ ,\ \})$ acts naturally on N_h by the Poisson bracket. We are interested by the Chevalley cohomology of this Lie algebra with values in N_h. In this case, a q-cochain C is an alternate multi-linear differential map of $(N_1)^q$ into N_h , the 0-cochains being identified with the elements of N_h. The coboundary of the q-cochain C is the $(q + 1)$-cochain ∂C given by :

$$\partial C(u_o,\dots,u_q) = \varepsilon_{o\dots q}^{\alpha_o\dots\alpha_q}\Big(\frac{1}{q!}\{u_{\alpha_o}, C(u_{\alpha_1},\dots,u_{\alpha_q})\}$$

$$- \frac{1}{2(q-1)!}\, C(\{u_{\alpha_o}, u_{\alpha_1}\}, u_{\alpha_2},\dots,u_{\alpha_q})\Big) \qquad (3\text{-}2)$$

where $u_\alpha \in N$, and where ε is the Kronecker skewsymmetrization indicator. We denote by $H^q(N_1;N_h)$ the qth cohomology space and by $H(N_1;N_h)$ the cohomology itself.

b) Let $\hat{N}_h$ be the space of the scalar densities of $\hat{W}$ with the weight $-h/(n+1)$. It is easy to prove that there is a canonical isomorphism ν of $\hat{N}_h$ onto N_h given by :

$$\nu : \bar{u} \in \hat{N}_h \to u = e^{hz^\circ}\, p^* \,\bar{u} \in N_h \qquad (3\text{-}3)$$

Compare the cohomology of N_1 with values in N_h and the cohomology of $\hat{L}$ with values in $\hat{N}_h$ on which $\hat{L}$ acts by Lie derivative. The coboundary $\hat{\partial}$ of a q-cochain $\hat{C}$ with values in $\hat{N}_h$ is given by:

$$\hat{\partial}\,\hat{C}(\hat{X}_o,\dots,\hat{X}_q) = \varepsilon_{o\dots q}^{\alpha_o\dots\alpha_q}\Big(\frac{1}{q!}\, L(\hat{X}_{\alpha_o})C(\hat{X}_{\alpha_1},\dots,\hat{X}_{\alpha_q})$$

$$- \frac{1}{2(q-1)!}\,\hat{C}\,([\hat{X}_{\alpha_o}, \hat{X}_{\alpha_1}], \hat{X}_{\alpha_2},\dots,\hat{X}_{\alpha_q})\Big) \qquad (3\text{-}4)$$

where $\hat{X}_\alpha \in \hat{L}$. The isomorphism between $\hat{L}$ and N_1 and a direct calculus on the coboundaries using the properties of the Lie derivative show :

Proposition.- ν gives a canonical isomorphism between the cohomology $H(\hat{L};\hat{N}_h)$ of $\hat{L}$ with values in $\hat{N}_h$ corresponding to the Lie derivative and the cohomology $H(N_1;N_h)$.

I have proved [3] that the space $\hat{N}_1$ is canonically isomorphic to $\hat{L}$. The corresponding bracket on $\hat{N}_1$ is given by the action on $\hat{N}_1$ of the Lie derivative by the image of an element of $\hat{N}_1$ in $\hat{L}$. Therefore for $h = 1$, the cohomologies $H(\hat{L};\hat{L})$, $H(\hat{L};\hat{N}_1)$ and $H(N_1;N_1)$ are canonically isomorphic.

For $h = 0$, $H(\hat{L};\hat{N})$, where $\hat{L}$ acts on $\hat{N}$ by the Lie derivative, is canonically isomorphic to $H(N_1;N_0)$.

We will see that the case $h = 1$ is particularly interesting.

c) We will be led to consider on W functions with complex values, elements of $N^c = C^\infty(W;C)$. We denote by N_h^c $(h \in \mathbb{R})$ the subspace of N^c defined by the elements which are homogeneous of degree h. It follows trivially from (3-2) that we have for the cohomology of the Lie algebra N_1, with values in N_h^c ,corresponding to the Poisson bracket :

$$H^q(N_1;N_h^c) \simeq H^q(N_1;N_h) \times H^q(N_1;N_h)$$

4.-Pseudodifferential operators and cohomologies.

We recall briefly the results of Omori [4] concerning the pseudo-differential operators of order 1 on a differentiable manifold M of dimension $(n + 1) \geq 2$.

a) Let $W = T_o^*M$ be the cotangent bundle of M without its null section and $\hat{W} = \Theta^*M$ the bundle of the cotangent oriented rays for M . On $W = T_o^*M$, the Liouville 1-form ω determines a structure of exact symplectic manifold which defines on $\hat{W} = \Theta^*M$ by quotient a contact structure $\hat{\pi}$. The exact symplectic manifold associated with $(\hat{W},\hat{\pi})$ can be identified with $(W = T_o^*M,\omega)$.

By means of the choice of a riemannian metric of M , Omori gives a representation of a class of invertible Fourier integral operators on M and of pseudodifferential operators with homogeneous symbols. In a suitable sense, the Lie algebra of the group corresponding to the Fourier integral operators of order 0 is given by ip^1, where p^1 is the space of the pseudodifferential operators of order 1, with real principal symbols [4].

b) More generally, let p^h $(h \in Z)$ be the space of all the pseudo-differential operators of order h. It is well known that $p^{-\ell}$ $(\ell \geq 0)$ is a Lie ideal of p^1 and that $(p^{-\ell}/p^{-\ell-1})$ is naturally isomorphic to the abelian Lie algebra given by $N_{-\ell}^c$. In particular p^o/p^{-1} is isomorphic with the abelian algebra N_o^c . On the other hand, it is possible to prove that p^1/p^o is naturally isomorphic with the Lie algebra $(N_1,\{\ ,\ \})$ and we have the exact sequence of Lie algebras

$$0 \to p^{o}/p^{-1} \simeq N_o^c \to p^1/p^{-1} \to p^1/p^o \simeq N_1 \to 0 \qquad (4\text{-}1)$$

p^0/p^{-1} being abelian, this sequence defines a cohomology 2-class, element of $H^2(N_1;N_o^c)$.

Omori proves by a long argument using the auxiliary riemannian structure that this class vanish always and the sequence (4-1) splits.

It follows that there is a subalgebra A of p^1 , which contains p^{-1}, such that A/p^{-1} is isomorphic with p^1/p^0 . We have the exact sequence :

$$0 \to p^{-1}/p^{-2} \simeq N_{-1}^c \to A/p^{-2} \to A/p^{-1} \simeq N^1 \to 0 \qquad (4\text{-}2)$$

p^{-1}/p^{-2} being abelian, this sequence defines also a cohomology 2-class β_o , element of $H^2(N_1;N_{-1}^c)$. Omori has proved by a long argument that β_o is always $\neq 0$ (theorem B of [4]).

We will study <u>for an arbitrary contact manifold</u>, the two spaces $H^2(N_1;N_o)$ and $H^2(N_1;N_{-1})$ and we will see how the previous results appear in this general framework.

5.- 1-differential cohomology $H_{(1)}(\hat{L};\hat{N}_h)$

a) A cochain $\hat{C}$ of $\hat{L}$ (or $\hat{N}_1$), with values in $\hat{N}_h$, is said to be 1-differential if it is given by a multidifferential operator on $\hat{N}_1$ of maximum order 1 in each argument. It is easy to see that if $\hat{C}$ is 1-differential, its coboundary is also 1-differential. We are thus led to study the cohomology $H_{(1)}(\hat{L};\hat{N}_h)$ corresponding to 1-differential cochains.

We can prove that this cohomology is canonically isomorphic with the 1-differential cohomology $H_{(1)}(N_1;N_h)$. But we know that, on a symplectic manifold, the coboundary of a 1-differential cochain C of the Poisson Lie algebra $(N;\{\ ,\ \})$, with values in N , which is defined by a q-tensor denoted again by C , is given by :

$$\mu(\partial C) = - d\mu(C) \qquad (5\text{-}1)$$

Choose an auxiliary pfaffian structure on $\hat{W}$. With such a choice, a q-cochain C corresponding to $H_{(1)}(N_1;N_h)$ can be described by a pair $(\hat{\alpha},\hat{\beta})$ of a q-form and a (q - 1)-form of $\hat{W}$. It follows from (5-1) that :

$$\partial(\hat{\alpha},\hat{\beta}) = (d\hat{\alpha},\ h\hat{\alpha} - d\hat{\beta}) \tag{5-2}$$

$\hat{C}$ is a q-cocycle iff $d\hat{\alpha} = 0$, $h\hat{\alpha} - d\hat{\beta} = 0$; this cocycle is exact iff there is a (q-1)-cochain corresponding to $(\hat{\alpha}',\hat{\beta}')$ such that $\hat{\alpha} = \hat{d}\hat{\alpha}'$, $\hat{\beta} = h\hat{\alpha}' - d\hat{\beta}'$. If $h \neq 0$ each q-cocycle is exact $(\hat{\alpha}' = \hat{\beta}/h,\ \hat{\beta}' = 0)$.

If $h = 0$, we have a cocycle if $\hat{\alpha}$ and $\hat{\beta}$ are closed and this cocycle is exact if $\hat{\alpha},\hat{\beta}$ are exact.

Theorem 1.- For $h \neq 0$, each cohomology space $H^q_{(1)}(N_1;N_h)$ or $H^q_{(1)}(\hat{L};\hat{N}_h)$ is null. For $h = 0$, each cohomology space $H^q_{(1)}(N_1;N_o)$ or $H^q_{(1)}(\hat{L};\hat{N})$ is isomorphic with the product of the De Rham cohomology spaces $H^q(\hat{W};\mathbb{R}) \times H^{q-1}(\hat{W};\mathbb{R})$ and has the dimension :

$$b_q(\hat{W}) + b_{q-1}(\hat{W})$$

where $b_q(\hat{W})$ is the qth Betti number of $\hat{W}$ for the cohomology with unrestricted supports.

b) We can study by means of a study of the differential types that a differential 1-cocycle $\hat{C}$ of $\hat{L}$ with values in $\hat{N}_h$ is necessarily 1-differential. It follows from Theorem 1 :

Theorem 2.- For $h \neq 0$, the cohomology spaces $H^1(N_1;N_h)$ or $H^1(\hat{L};\hat{N}_h)$ are null. For $h = 0$, the cohomology spaces $H^1(N_1;N_0)$ or $H^1(\hat{L};\hat{N})$ are isomorphic with $\mathbb{R} \times H^1(\hat{W};\mathbb{R})$ and have the dimension $1 + b_1(\hat{W})$.

Let $\hat{k}$ be a positive kernel. It is easy to see that $\hat{X} \in \hat{L} \to L(\hat{X})\hat{k}/\hat{k} = L(\hat{X}) \log\hat{k}$ defines a 1-cocycle of $\hat{L}$ with values in $\hat{N}$. The corresponding cohomology 1-class is independent of the choice of $\hat{k}$ and is always $\neq 0$. Each 1-cocycle $\hat{C}$ of $\hat{L}$ with values in $\hat{N}$ is given by :

$$\hat{C}(\hat{X}) = c\, L(\hat{X}) \log\hat{k} + \hat{\alpha}(\hat{X}) \qquad (\hat{X} \in \hat{L};\ c \in \mathbb{R}) \tag{5-3}$$

where $\hat{\alpha}$ is an arbitrary closed 1-form of $\hat{W}$.

6.- Differential cohomology in dimension two.

a) It is possible to prove by a long study of the bidifferential types that, for $h \neq -1$, each differential 2-cocycle of $\hat{L}$ with values in $\hat{N}_h$ is cohomologous to a 1-differential 2-cocycle. It follows :

Theorem 3.- For $h \neq -1, 0$ the cohomology spaces $H^2(N_1;N_h)$ or $H^2(\hat{L};\hat{N}_h)$ are null. For $h = 0$, the cohomology spaces $H^2(N_1;N_o)$ or $H^2(\hat{L};\hat{N})$ have the dimension $b_1(\hat{W}) + b_2(\hat{W})$.

With notations similar to (5-3), each 1-differential 2-cocycle $\hat{C}$ of $\hat{L}$ with values in $\hat{N}$ is given by :

$$\hat{C}(\hat{X};\hat{Y}) = \hat{\alpha}(\hat{X};\hat{Y}) + (\hat{\beta}(\hat{X})L(\hat{Y}) \log \hat{k} - \hat{\beta}(\hat{Y})L(\hat{X}) \log \hat{k} \quad (\hat{X},\hat{Y} \in \hat{L}) \tag{6-1}$$

where $\hat{\alpha}$ (resp. $\hat{\beta}$) is an arbitrary closed 2-form (resp. 1-form) of $\hat{W}$.

For $h = 1$, we deduce from Theorem 3 and from the theory of deformations (see [5] ,[7]).

Corollary.- Each differential deformation of the Lie algebra $\hat{L}$ of the infinitesimal automorphisms of a contact manifold is differentiably trivial.

This Lie algebra is thus rigid. I have proved previously this rigidity [7].

b) Study the case $h = -1$, which is the most interesting case. Let Γ be a symplectic connection on a symplectic manifold (W,F). If $u \in N$, we denote by $L(X_u)\Gamma$ the Lie derivative of Γ by the hamiltonian vector field defined by u. We know, [5], [10] that the 2-cochain of the Poisson Lie algebra $(N,\{\ ,\ \})$ with values in N which is given in each domain U of an arbitrary chart $\{x^A\}$ of W by :

$$S^3_\Gamma(u,v)\big|_U = - \Lambda^{AB}(L(X_u)\Gamma)^R_{SA}\ (L(X_v)\Gamma)^S_{RB} \tag{6-2}$$

is a Chevalley 2-cocycle of bidifferential type (3,3). Its cohomology class is independent of the choice of the connection and is always $\neq 0$. This 2-class is an invariant of the symplectic structure of (W,F).

We have seen that, for the exact symplectic manifold (W,ω) associated with $(\hat{W},\hat{\pi})$, there are symplectic connections invariant under Z. For such a choice of the connection Γ, the restriction of S^3_Γ to N_1 is a 2-cocycle of $(N_1,\{\ ,\ \})$ with values in N_{-1} and it is easy to see that its cohomology 2-class β_1, element of $H^2(N_1;N_{-1})$ is independent of the choice of the connection.

On the other hand, we can prove by a standard argument [1] :

Lemma.- If T is a cochain of N_1, with values in N_{-1} such that its coboundary is d-differential $(d \geq 1)$, T is necessarily a differential operator of order d.

It follows that if the restriction of S^3_Γ to N_1 is the coboundary of an arbitrary 1-cochain, it is the coboundary of a differential operator. It is then possible to prove by means of the introduction of two pairs of suitable functions, elements of N_1 [4], with proportional Poisson brackets :

Proposition.- The cohomology class β_1, element of $H^2(N_1;N_{-1})$ or of $H^2(\hat{L};\hat{N}_{-1})$, defined by means of S^3_Γ is always $\neq 0$. It is an invariant of the contact manifold $(\hat{W},\hat{\pi})$.

The same argument as in the case $h \neq -1$ proves :

Theorem 4.- The cohomology space $H^2(N_1;N_{-1})$ admits the unique generator $\beta_1 \neq 0$ defined by S^3_Γ. We have always $\dim H^2(N_1;N_{-1}) = \dim H^2(\hat{L};\hat{N}_{-1}) = 1$.

Our introduction [6] by Flato and myself of the cohomology of the Lie algebra of all vector fields of a manifold associated to the Lie derivative of forms and the corresponding study of M de Wilde and P. Lecomte [8] lead to the consideration of the 2q-cochains $S^{3,q}_\Gamma$ of N_1 with values in N_{-q} given by :

$$S^{3,q}_\Gamma(u_1,\ldots,u_{2q}) = -\Lambda^{[A_1A_2}\Lambda^{A_3A_4}\ldots\Lambda^{A_{2q-1}A_{2q}]} \tag{6-3}$$

$$(L(X_{u_1})\Gamma)^{R_1}_{R_2A_1}(L(X_{u_2})\Gamma)^{R_2}_{R_3A_2}\ldots(L(X_{u_{2q}})\Gamma)^{R_{2q}}_{R_1A_{2q}}$$

where $u_\alpha \in N_1$ and where $[\ ,\ ]$ denotes the skewsymmetrization. The $S_\Gamma^{3,q}$ are 2q-cocycles of N_1 with values in N_{-q} and the corresponding cohomology classes are independent of the choice of Γ. We can conjecture that (6-3) gives generators of $H^{2q}(N_1;N_{-q})$.

We note that we have proved that S_Γ^3 is never the coboundary of a 1-cochain, without assumption of locality, continuity or differentiability on this 1-cochain. Therefore β_1 can be considered as an element of a general cohomology space.

c) In the particular case $W = T_o^*M$ considered by Omori, it is possible to prove that the de Rham cohomology corresponding to $h = 0$ does not interfere in the exact sequence (4-1) and that we have :

Proposition.- The cohomology 2-class β_o associated with the exact sequence (4-2) is connected with the invariant β_1 corresponding to $\hat{W} = 0^*M$ by :

$$\beta_o = (1/4)\ \beta_1 \tag{6-4}$$

We see how the Omori Theorem B is a particular case of our theorem 4.

REFERENCES

[1] A. Lichnerowicz. J. Math.pures et appl. t57, 1978, p 453-488.

[2] V. Arnold. Méthodes mathématiques de la Mécanique classique. Mir Moscou 1976, p 358-361.

[3] A. Lichnerowicz. C.R. Acad. Sci. Paris t 290 A, 1980, p 963; t 291 A, 1980, p 129.

[4] H. Omori, Y. Maeda, A. Yohioka, O. Kobayashi. On regular Frechet-Lie groups (III) : a second cohomology class related to the Lie algebra of pseudodifferential operators of order one. Tokyo J.of Math. (to appear).

[5] J. Vey. Comm. Math. Helv. t 50, 1975, p 421-454.

[6] M. Flato and A. Lichnerowicz. C. R. Acad. Sci. Paris t 291 A, 1980, p 331-335.

[7] A. Lichnerowicz. C.R. Acad. Sci. Paris t 290 A,1980, p 241-245.

[8] M. de Wilde and P. Lecomte. Cohomology of the Lie algebra of smooth vector fields of a manifold associated to the Lie derivative of smooth forms. J. Math. pures at appl. (to appear).

[9] M.V. Losik. Funct. Anal. and Appl. t 4, 1970, p 127-135.

[10] F.Bayen, M. Flato, C. Fronsdal, A. Lichnerowicz, D. Sternheimer. Deformation theory and quantization. Ann. of Physics, t 111, 1978, p 61-110.

[11] L. Hörmander. Acta Math. t 127, 1971, p 79-183.

[12] T. Ratiu and R. Schmid . Math.Zeit. t 177, 1981, p 81-100.

A NOTE ON SEMISIMPLE FLAT HOMOGENEOUS SPACES

J.F.T. Lopera
Dto. de Geometría y Topología
Facultad de Matemáticas
Universidad de Santiago de Compostela
Spain

ABSTRACT

Let $L_\circ$ be the isotropy group of the origin in a semisimple flat homogeneous space $M = L/L_\circ$ associated with a semisimple graded transitive Lie algebra $\ell = g_{-1} \oplus g_0 \oplus g_1$. It is a well known fact, due to T. Ochiai, that $L_\circ$ can be considered as a Lie subgroup of $G^2(m)$, the Lie group of 2-jets $j^2_0(g)$ at $0 \in g_{-1}$, where g is a diffeomorphism from a neighbourhood of $0 \in g_{-1} \equiv \mathbb{R}^m$ onto a neighbourhood of $0 \in g_{-1}$. The aim of this note is to show that $L_\circ$ is in fact a closed subgroup when it has only a finite number of connected components, giving an affirmative answer to a conjecture suggested to us by M. Takeuchi one year ago.

1.- PRELIMINARY FACTS.

Let us recall, [7], that a connected homogeneous space M is said to be semisimple flat associated to a semisimple graded (transitive and real) Lie algebra $\ell = g_{-1} \oplus g_0 \oplus g_1$ when there is a Lie group L acting transitively and effectively on M, and the isotropy group $L_\circ$ of the origin in $M = L/L_\circ$ having Lie algebra $\ell_\circ = g_0 \oplus g_1$. Since g_{-1} is an abelian Lie algebra, we shall identify it with $\mathbb{R}^m$, $m = \dim M$.

Let $G_\circ = N(g_0, L) \cap L_\circ$ be the normalizer of g_0 in $L_\circ$; then every $a \in L_\circ$ can only be expressed in one way as a product

$$a = g \exp Z \tag{1}$$

with $g \in G_\circ$ and $Z \in g_1$. If B denotes the Killing form of ℓ, then the map

$$Z \in g_1 \longmapsto B(-,Z)\big|_{g_{-1}} \in g_{-1}^* \tag{2}$$

defines a canonical isomorphism of g_1 onto g_{-1}^*.

The so called linear isotropy representation of $L_\circ$

$$l: a \in L_\circ \longrightarrow l(a) \in GL(g_{-1}) \tag{3}$$

given as

$$l(a)X = Ad_L(a)X \quad \text{modulo}(\ell_\circ) \tag{4}$$

is faithful on $G_\circ$; in this way, we can identify $G_\circ$ with the linear isotropy group of $M = L/L_\circ$.

Let us recall the following important theorem, Ochiai [7]:

THEOREM

The homomorphism $j: L_\circ \longrightarrow G^2(m)$ given by

$$j(a) = j_0^2(\mathrm{Exp}^{-1} \circ \tau_a \circ \mathrm{Exp}) \tag{5}$$

is injective, where $\mathrm{Exp}: X \in g_{-1} \longrightarrow (\exp X)L_\circ \in M$ defines a local chart at the origin of M, and $\tau_a(xL_\circ) = (ax)L_\circ$, for every $xL_\circ$ M.

2.- A FORMULA TO COMPUTE THE ADJOINT REPRESENTATION OF ORDER TWO. SOME CONSEQUENCES.

Let P^rM be the $G^r(m)$ bundle of frames of order r on a manifold M, $\dim(M) = m$. We recall, [2], that for every $j_0^2 g \in G^2(m)$ there is a local diffeomorphism of $(P^1\mathbb{R}^m, j_0^2(id))$ defined by

$$j_0^1 f \longmapsto j_0^1(g \circ f \circ g^{-1}). \tag{6}$$

Its differential at $e = j_0^1(id)$ is a linear transformation of $T_eP^1\mathbb{R}^m \equiv ga(\mathbb{R}^m) \equiv \mathbb{R}^m \oplus g\ell(\mathbb{R}^m)$ which will be denoted $Ad^{(2)}(j_0^2 g)$. The linear representation of $G^2(m)$ in $ga(\mathbb{R}^m)$ given by $Ad^{(2)}$ is called the second order adjoint representation of $G^2(m)$.

As far as we know, the following explicit formula to compute $Ad^{(2)}(j_0^2 g)$ has not been developed before. It will play a key role in this paper.

PROPOSITION 1.-

If $j_o^2(g) \in G^2(m)$ is given as the pair $(\dot{g},\ddot{g})$, with $\dot{g} = Dg(0)$, $\ddot{g} = D^2g(0)$, then for every $(x,A) \in \mathbb{R}^m \oplus g\ell(\mathbb{R}^m)$ we have

$$\mathrm{Ad}^{(2)}(j_0^2 g)\begin{pmatrix} x \\ A \end{pmatrix} = \begin{pmatrix} \dot{g}(-) & 0 \\ \bar{\ddot{g}}(-)g^{-1} & \dot{g}(-)\dot{g}^{-1} \end{pmatrix}\begin{pmatrix} x \\ A \end{pmatrix} \tag{7}$$

where for every bilinear map α

$$\bar{\alpha}(x)(y) = \alpha(x,y). \tag{8}$$

Proof

It follows by differentiation of an appropiate decomposition of the map given in (6). For details see [5]. #

As a first consequence of this proposition we obtain the following characterization of $j(G_\circ)$

PROPOSITION 2.-

$$j(G_\circ) = \{j(a) \in jL_\circ;\ \ \mathrm{Ad}^{(2)}(ja)\,\mathbb{R}^m \subset \mathbb{R}^m\} \tag{9}$$

Proof

It follows from (1), (7) and the theorem of Ochiai in §1. For details see [5]. #

Now, if $a \in L_\circ$ then $\mathrm{Ad}_L(a)\,\mathfrak{g}_1 \subset \mathfrak{g}_1$, and we obtain an induced map

$$\widetilde{\mathrm{Ad}_L}(a) : \ell/\mathfrak{g}_1 \equiv \mathfrak{g}_{-1} \oplus \mathfrak{g}_0 \longrightarrow \ell/\mathfrak{g}_1 \equiv \mathfrak{g}_{-1} \oplus \mathfrak{g}_0 \tag{10}$$

Moreover, $\mathfrak{g}_0$ can be considered as a Lie subalgebra of $\mathfrak{gl}(\mathfrak{g}_{-1})$ by way of the linear isotropy representation

$$\lambda : Y \in \mathfrak{g}_0 \longrightarrow [Y,-] \in \mathfrak{gl}(\mathfrak{g}_{-1}) \tag{11}$$

Then, the following proposition of Ochiai [7] follows, (see also [2] and [9]).

PROPOSITION 3.-

If $a \in L_\circ$ then

i) $\mathrm{Ad}^{(2)}(ja)(\mathfrak{g}_{-1} \oplus \mathfrak{g}_0) \subset \mathfrak{g}_{-1} \oplus \mathfrak{g}_0$

ii) $\mathrm{Ad}^{(2)}(ja)\big|_{\mathfrak{g}_{-1} \oplus \mathfrak{g}_0} = \widetilde{\mathrm{Ad}_L}(a)$ #

Let G_1 be the connected Lie subgroup of L with Lie algebra g_1. It is well known, [7], [8], that the exponential map from g_1 into G_1 is bijective.

PROPOSITION 4.-

$j(G_1)$ is a closed subgroup of $G^2(m)$.

Proof

Using Propositions 1 and 3, and bearing in mind that

$$Ad_L(\exp Z)(X) = X + [Z,X] \tag{12}$$

for any $Z \in g_1$ and $X \in g_{-1}$, one can deduce that

$$\dot{g} = id \quad , \quad \ddot{g}(X) = \lambda([Z,X]) \tag{13}$$

where $j_0^2(g) = j(\exp Z)$. Therefore

$$j(G_1) = \{id\} \times V \subset G^2(m) \tag{14}$$

where V denotes the linear space of symmetric bilinear maps from $g_{-1} \times g_{-1}$ into g_{-1} given as

$$[[Z,-],-] \tag{15}$$

for each $Z \in g_1$ #

PROPOSITION 5.-

If $l(G_\circ)$ is a closed subgroup of $GL(g_{-1})$ then $j(L_\circ)$ is a closed subgroup of $G^2(m)$.

Proof

Let $\{a_n = g_n \exp Z_n \ ; \ n \in \mathbb{N}\}$ be a sequence in $L_\circ$, see (1), such that $\{j(a_n) \ ; \ n \in \mathbb{N}\}$ converges to $j_0^2 f = (B,\beta) \in G^2(m)$.
Let us define

$$f_n = \mathrm{Exp}^{-1} \circ \tau_{g_n \exp Z_n} \circ \mathrm{Exp} \tag{16}$$

$$\phi_n = \mathrm{Exp}^{-1} \circ \tau_{g_n} \circ \mathrm{Exp} \tag{17}$$

$$\psi_n = \mathrm{Exp}^{-1} \circ \tau_{\exp Z_n} \circ \mathrm{Exp} \tag{18}$$

Then $f_n = \phi_n \circ \psi_n$ and taking into account (13) and Proposition 2 we arrive to

$$Df_n(0) = D\phi_n(0) \tag{19}$$

$$D^2f_n(0) = D\phi_n(0)\circ D^2\psi_n(0). \tag{20}$$

By hypothesis

$$\lim_{n\to\infty} D\phi_n(0) = B. \tag{21}$$

Since $l(G_o) \equiv j(G_o)$ is closed, then there is a unique element $x \in G_o$ such that $(B,0) = j(x)$. Moreover, we have

$$B^{-1}\circ\beta = \lim_{n\to\infty} D^2\psi_n(0) \tag{22}$$

by virtue of (18), (19), (20), (21) and proposition 4; then

$$(I, B^{-1}\circ\beta) \in j(G_1) \quad , \quad (I = id_{\mathbb{R}^m}). \tag{23}$$

Finally we obtain

$$(B,\beta) = (B,0)(I,B^{-1}\circ\beta) \in j(L_o). \qquad \#$$

Actually, we are uncertain whether the hypothesis "$l(G_o)$ is closed in $GL(g_{-1})$" is necessary in the first place; but, in any case, in the following examples this hypothesis is always verified:

1) $M = P(\mathbb{R}^n)$, the projective space, $L = PGL(\mathbb{R}^n)$, the projective group, $\ell = s\ell((n-1)+1;\mathbb{R})$, see [2], [6].

2) $M = E_{p,q} = (S^p \times S^q)/\sim$, where $(x,y)\sim(x',y')$ if and only if $x' = -x$, $y' = -y$. $L = O(S)/\{\pm I\}$, $\ell = o(S)$, with

$$S = \begin{pmatrix} 0 & 0 & -1 \\ 0 & \begin{matrix} I_p & \\ & I_q \end{matrix} & 0 \\ -1 & 0 & 0 \end{pmatrix} \tag{24}$$

and $O(S)$ the orthogonal group associated with S; see [6], [9].

3) $M = G_p(\mathbb{R}^n)$, the Grassmann manifold of p-planes in $\mathbb{R}^n$, $0<p<n$, $L = PGL(\mathbb{R}^n)$, $\ell = s\ell(p+q,\mathbb{R})$, see [5]. This example also works in the complex case (see [5,III §4]) and it falls within the scope of the complex theory developped by Ochiai [8]. Note that example 1 is $G_1(\mathbb{R}^n)$.

3.- MAIN RESULT.

Let us define a bilinear map, Tanaka [10],

$$\Psi: g_{-1}\times g_1 \longrightarrow gl(g_{-1}) \tag{25}$$

by setting

$$\Psi(\xi,\phi) = \lambda[\xi,\phi] \quad , \quad \xi\in g_{-1}, \phi\in g_1 \tag{26}$$

where λ is given by (11). Then, by virtue of (2), Ψ can be considered as a tensor of type (2,2). Moreover, the identity representation of $GL(g_{-1})$ in g_{-1} induces a canonical action of $GL(g_{-1})$ on the tensor product $(\overset{r}{\otimes} g_{-1}) \otimes (\overset{s}{\otimes} g_{-1}{}^*)$. Then

PROPOSITION 6.-

Every $a\in L_\circ$ leaves Ψ invariant through the linear isotropy representation (4).

Proof

Direct computation. #

Let G_Ψ be the isotropy group of tensor Ψ under the action of $GL(g_{-1})$; G_Ψ is a closed subgroup of $GL(g_{-1})$ and, by virtue of Proposition 6, it contains $l(L_\circ) = l(G_\circ) = j(G_\circ)$. Therefore, if g_Ψ denotes the Lie algebra of G_Ψ then $\lambda(g_0)\subset g_\Psi$. It is not difficult to show that an element $x\in GL(g_{-1})$ leaves Ψ invariant if and only if

$$\Psi(x(\xi),\phi\circ x^{-1}) = x\circ\Psi(\xi,\phi)\circ x^{-1} \quad , \quad \xi\in g_{-1}, \quad \phi\in g_{-1}{}^* \tag{27}$$

and that if an element $A\in gl(g_{-1})$ belongs to g_Ψ then it follows that, for every $\xi\in g_{-1}$, $\phi\in g_{-1}{}^*$

$$A\circ\Psi(\xi,\phi) - \Psi(\xi,\phi)\circ A = \Psi(A\xi,\phi) - \Psi(\xi,\phi\circ A) \tag{28}$$

These A satisfying (28) form in fact a Lie subalgebra g of $gl(g_{-1})$

PROPOSITION 7.-

$\lambda(g_0) = g = g_\Psi$

Proof

See [1], or [5] for a variation on this proof. #

PROPOSITION 8.-

Let $\tilde{G}_o$ and $\tilde{G}_\psi$ be the connected component of the identity in G_o and G_ψ respectively. Then $l(\tilde{G}_o) = \tilde{G}_\psi$.

Proof

It follows directly from Proposition 7. #

Ochiai, [8], has proved an analogous statement, suppossing ℓ to be simple, that is to say, λ irreducible, (11), see Tanaka [10].

We must recall that G_o is a strong deformation retract of L_o; the homotopy

$$(g\exp Z, t) \in L_o \times [0,1] \longmapsto g\exp(tZ) \in L_o \qquad (29)$$

gives such a deformation. Consequently G_o and L_o have the same number of connected components. Then we can state

PROPOSITION 9.-

If G_o (or equivalently L_o) has a finite number of connected components then $l(G_o)$ is closed in $GL(g_{-1})$ and thus $j(L_o)$ is a closed subgroup of $G^2(m)$.

Proof

$l(G_o) = G_\psi$ is a closed subgroup of $GL(g_{-1})$. Now if we choose elements $x_1, \ldots, x_k$, each one of them in a different component of G_o, we get

$$G_o = x_1 G_o \cup \ldots \cup x_k G_o \qquad (30)$$

and then

$$l(G_o) = l(x_1) l(G_o) \cup \ldots \cup l(x_k) l(G_o) \qquad (31)$$

which is a finite union of closed subsets of $GL(g_{-1})$. Lastly, taking into account Proposition 5, we reach our final conclusion. #

Let us conclude by pointing out that if $j(L_o)$ is closed then $j(L_o)$-structures on a manifold M of the same dimension that L/L_o can be considered as global cross-sections $\sigma: M \longrightarrow P^2M/j(L_o)$. On the basis of this fact, we have studied the theory of equivalent connections adapted to a $j(G_0)$-structure on M, ([5], Proposition IV.3-7).

BIBLIOGRAPHY

[1] T. Hangan, 1-systèmes de N. Tanaka et structures tensorielles associées. Journées franco-belges de géometrie différentielle (1978)

[2] S. Kobayashi and T. Nagano, On projective connections. J. Math. Mech. 13,(1964),215-235.

[3] S. Kobayashi and T. Nagano, On filtered Lie algebras and geometric structures. I, J. Math. Mech. 13,(1964),875-908; II Ibid. 14,(1965),516-522.

[4] S. Kobayashi and T. Ochiai, G-structures of order two and transgression operators. J. Diff. Geom. 6,(1971),213-230.

[5] J.F.T. Lopera, Espacios homogéneos semisimples llanos, estructuras grassmannianas y foliaciones transversalmente grassmannianas. Publ. Departamento de Geomatría y Topología, n°57. Univ. De Santiago de Compostela (1982).

[6] S. Nishikawa and M. Takeuchi, Γ-foliations and semisimple flat homogeneous spaces, Tôhoku Math. J. 30,(1978),307-335.

[7] T. Ochiai, Geometry associated with semisimple flat homogeneous spaces. Trans. Amer. Math. Soc. 152,(1970),159-193.

[8] T. Ochiai, A survey on holomorphic G-structures. Preprint.

[9] K. Ogiue, Theory of conformal connections. Kodai Math. Sem. Report 19,(1967),193-224.

[10] N. Tanaka, On the equivalence problems associated with certain class of homogeneous spaces. J. Math. Soc. Japan, 17,(1965), 103-139.

SOME RESULTS ON $DIFF^{\Omega}(\mathbb{R}^n)$

F. Mascaró
Dept. Geometría y Topología
Fac. Matemáticas. Valencia. SPAIN.

We denote by Ω any volume element on $\mathbb{R}^n$ and by $Diff^{\Omega}(\mathbb{R}^n)$ the group of all diffeomorphisms of $\mathbb{R}^n$ preserving Ω. In [1] we gave a complete classification of the normal subgroups of $Diff^{\Omega}(\mathbb{R}^n)$ for $n \geqslant 3$ and Ω a volume element of finite total volume. Thus, we get that a subgroup N of $Diff^{\Omega}(\mathbb{R}^n)$ is normal if and only if

$$Diff^{\Omega}_{co}(\mathbb{R}^n) \subset N \subset Diff^{\Omega}_{c}(\mathbb{R}^n)$$

where $Diff^{\Omega}_{co}(\mathbb{R}^n)$ is the normal subgroup of $Diff^{\Omega}(\mathbb{R}^n)$ of all elements compactly isotopic to the identity by an isotopy preserving Ω and $Diff^{\Omega}_{c}(\mathbb{R}^n)$ is the normal subgroup of $Diff^{\Omega}(\mathbb{R}^n)$ of all elements with compact support.

When Ω has infinite total volume we obtained the following chain of normal subgroups of $Diff^{\Omega}(\mathbb{R}^n)$ (see [1], [2])

$$\{id\} \text{ —— } Diff^{\Omega}_{co}(\mathbb{R}^n) \subset Diff^{\Omega}_{c}(\mathbb{R}^n) \text{ —— } Diff^{\Omega}_{f}(\mathbb{R}^n) \subset$$
$$\subset Diff^{\Omega}_{W}(\mathbb{R}^n) \text{ —— } Diff^{\Omega}(\mathbb{R}^n)$$

where $Diff^{\Omega}_{f}(\mathbb{R}^n)$ is the normal subgroup of $Diff^{\Omega}(\mathbb{R}^n)$ of all elements with support of finite Ω-volume and $Diff^{\Omega}_{W}(\mathbb{R}^n)$ the normal subgroup of $Diff^{\Omega}(\mathbb{R}^n)$ of all elements with set of non-fixed points ($\text{fix}\, h = \{x \in \mathbb{R}^n : h(x) = x\}$) of finite Ω-volume. —— denotes that there is no normal subgroup in between.

To obtain a similar result to the case of finite total volume it remains to study the subgroups of $Diff^{\Omega}(\mathbb{R}^n)$ lying between $Diff^{\Omega}_{f}(\mathbb{R}^n)$ and $Diff^{\Omega}_{W}(\mathbb{R}^n)$.

In this direction, in §1, we study the closures of the subgroups of the above chain with respect to two topologies that make $\mathrm{Diff}^{\Omega}(\mathbb{R}^n)$ a topological group. Thus, we get that $\mathrm{Diff}^{\Omega}_{co}(\mathbb{R}^n)$ is dense with respect to the compact-open C^{∞}-topology. Also, we get that $\mathrm{Diff}^{\Omega}_{c}(\mathbb{R}^n)$ and $\mathrm{Diff}^{\Omega}_{W}(\mathbb{R}^n)$ are both closed with respect to the Whitney C^{∞}-topology. It remains to identify the closures of $\mathrm{Diff}^{\Omega}_{co}(\mathbb{R}^n)$ and of $\mathrm{Diff}^{\Omega}_{f}(\mathbb{R}^n)$ with respect to the last topology.

If Ω has infinite total volume we get in [1] also that any subgroup N of $\mathrm{Diff}^{\Omega}(\mathbb{R}^n)$ satisfying $\mathrm{Diff}^{\Omega}_{co}(\mathbb{R}^n) \subset N \subset \mathrm{Diff}^{\Omega}_{c}(\mathbb{R}^n)$ is normal. In §2 we prove that this result is not true for the subgroups lying between $\mathrm{Diff}^{\Omega}_{f}(\mathbb{R}^n)$ and $\mathrm{Diff}^{\Omega}_{W}(\mathbb{R}^n)$ by constructing a normal subgroup of $\mathrm{Diff}^{\Omega}_{W}(\mathbb{R}^n)$ containing $\mathrm{Diff}^{\Omega}_{f}(\mathbb{R}^n)$ that is not normal in $\mathrm{Diff}^{\Omega}(\mathbb{R}^n)$.

I would like to thank D. McDuff and R. Sivera for their valuable suggestions and comments.

§1.- $\mathrm{Diff}^{\Omega}(\mathbb{R}^n)$ as a topological group

It is well known that $\mathrm{Diff}^{\Omega}(\mathbb{R}^n)$ is a topological group with respect to the compact-open C^{∞}-topology and with respect to the Whitney C^{∞}-topology.

Let $\mathrm{Diff}^{\Omega}_{co}(\mathbb{R}^n)$ be the normal subgroup of $\mathrm{Diff}^{\Omega}(\mathbb{R}^n)$ of all elements compactly isotopic to the identity by an isotopy preserving Ω. We will prove

1.1 Proposition.- If $n \geqslant 3$ $\mathrm{Diff}^{\Omega}_{co}(\mathbb{R}^n)$ is dense in $\mathrm{Diff}^{\Omega}(\mathbb{R}^n)$ with respect to the compact-open C^{∞}-topology.

Proof. We will construct an element h lying in the closure of $\mathrm{Diff}^{\Omega}_{co}(\mathbb{R}^n)$ ($\mathrm{cl}\,\mathrm{Diff}^{\Omega}_{co}(\mathbb{R}^n)$) but not in $\mathrm{Diff}^{\Omega}_{W}(\mathbb{R}^n)$. Then, since in [1] we proved that there is no normal subgroup between $\mathrm{Diff}^{\Omega}_{W}(\mathbb{R}^n)$ and $\mathrm{Diff}^{\Omega}(\mathbb{R}^n)$ it must happen $\mathrm{cl}\,\mathrm{Diff}^{\Omega}_{co}(\mathbb{R}^n) = \mathrm{Diff}^{\Omega}(\mathbb{R}^n)$.

Let us construct h. Let $\{C_i\}_{i \geqslant 1}$ be the family of closed balls of $\mathbb{R}^n$ of centre $(i,0,\dots,0)$ and radius $1/4$. Let $\psi_i : \mathbb{R} \longrightarrow [0,1]$

be a bump function such that $\psi_i(r) = 0$ if either $-\infty < r \leqslant i-\frac{1}{4}$ or $i+\frac{1}{4} \leqslant r < +\infty$. For any $r \in \mathbb{R}$, we can define the matrix $M_i(r)$ as follows.

$$M_i(r) = \begin{pmatrix} \cos\psi_i(r) & -\sin\psi_i(r) & 0 \\ \sin\psi_i(r) & \cos\psi_i(r) & \\ & 0 & \\ & & I \end{pmatrix}$$

Thus, the map $h_i\colon \mathbb{R}^n \longrightarrow \mathbb{R}^n$ given by $h_i(x) = x\cdot M_i(\|x\|)$ is a volume preserving diffeomorphism with support in C_i. Furthermore, there exits, $\psi_i^t\colon \mathbb{R} \longrightarrow [0,1]$, a C^∞-family of bump functions such that, for any t, $\psi_i^t(r) = 0$ if either $-\infty < r \leqslant i-\frac{1}{4}$ or $i+\frac{1}{4} \leqslant r < +\infty$, $\psi_i^0(r) = 0$ for any $r \in \mathbb{R}$ and ψ_i^1 equals ψ_i. The map

$$H_i\colon \mathbb{R}^n \times I \longrightarrow \mathbb{R}^n$$

given by $H_i(x,t) = x\cdot M_i^t(\|x\|)$, where $M_i^t(\|x\|)$ is a matrix as above but defined using ψ_i^t instead of ψ_i, is an Ω-isotopy from h_i to the identity with support in C_i. Therefore, h_i is an element of $\mathrm{Diff}^\Omega_{co}(\mathbb{R}^n)$.

Since, h_i has support in C_i, for any i, we can define a new volume preserving diffeomorphism of $\mathbb{R}^n$, $h = \ldots\circ h_2\circ h_1$. Clearly, we have

$$\mathbb{R}^n - \mathrm{fix}\,h = \coprod_{i\geqslant 1} (\mathrm{int}\, C_i - (i,0,\ldots,0)).$$

So

$$\mathrm{vol}_\Omega(\mathbb{R}^n - \mathrm{fix}\,h) = \mathrm{vol}_\Omega(\coprod_{i\geqslant 1} C_i) = \sum_{i\geqslant 1} \mathrm{vol}_\Omega C_i = \infty.$$

Therefore, h does not lie in $\mathrm{Diff}^\Omega_W(\mathbb{R}^n)$.

On the other hand, h is the limit of the following sequence of volume preserving diffeomorphisms. $\{h_j\circ h_{j-1}\circ\ldots\circ h_1\}_{j\geqslant 1}$ with respect to the compact-open C^∞-topology. Since each element of the sequence lies in $\mathrm{Diff}^\Omega_{co}(\mathbb{R}^n)$, we have that h lies in the closure of $\mathrm{Diff}^\Omega_{co}(\mathbb{R}^n)$ with respect to the compact-open C^∞-topology.

As an immediate consequence we have

<u>1.2 Corollary</u>.- If $n \geqslant 3$, the closure of any normal subgroup of $\mathrm{Diff}^\Omega(\mathbb{R}^n)$ with respect to the compact-open C^∞-topology is the whole group $\mathrm{Diff}^\Omega(\mathbb{R}^n)$.

Let $\mathrm{Diff}^\Omega_c(\mathbb{R}^n)$ be the normal subgroup of $\mathrm{Diff}^\Omega(\mathbb{R}^n)$ of all elements with compact-support. We will prove

<u>1.3 Proposition</u>.- $\mathrm{Diff}_c^{\Omega}(\mathbb{R}^n)$ is closed in $\mathrm{Diff}^{\Omega}(\mathbb{R}^n)$ with respect to the Whitney C^{∞}-topology.

<u>Proof</u>.- Let h be any element of $\mathrm{Diff}^{\Omega}(\mathbb{R}^n)$ with non-compact support. We will construct a neighbourhood of h not intersecting $\mathrm{Diff}_c^{\Omega}(\mathbb{R}^n)$.

It is proved in [1], that, for any element $h \in \mathrm{Diff}^{\Omega}(\mathbb{R}^n)$ with non-compact support there is a locally finite sequence of disjoint closed balls, $\coprod_{i\geq 1} C_i$, such that

$$(\coprod_{i\geq 1} C_i) \cap (\coprod_{i\geq 1} h(C_i)) = \emptyset$$

Let x_i, r_i, be the centre and radius of the ball C_i respectively. We denote by C_i' the closed ball of centre x_i and radius $r_i/2$.

We define

$$N(h; \{\mathrm{int}\, h(C_i)\}_{i\in\mathbb{N}}, \{C_i'\}_{i\in\mathbb{N}}, \{r_i/2\}_{i\in\mathbb{N}}) =$$

$$= \{g \in \mathrm{Diff}^{\Omega}(\mathbb{R}^n) : g(C_i') \subset \mathrm{int}\, h(C_i),\ \|D^k(h)(x) - D^k(g)(x)\| < r_i/2$$

$$\text{for all } i \in \mathbb{N}, \text{ for any } x \in C_i' \text{ and any } k \geq 0\}$$

obviously, it is a neigbourhood of h in $\mathrm{Diff}^{\Omega}(\mathbb{R}^n)$ with the Whitney C^{∞}-topology.

It does not meet $\mathrm{Diff}_c^{\Omega}(\mathbb{R}^n)$ since if f is an element of $\mathrm{Diff}_c^{\Omega}(\mathbb{R}^n)$ there is some index $j \in \mathbb{N}$ such that $(\mathrm{supp}\, f) \cap C_j = \emptyset$. Therefore, for any point $x \in C_j'$ we have $h(x) \neq x$ and $f(x) = x$, thus, $\|x - h(x)\| > r_j/2$. So, f is not an element of $N(h; \{\mathrm{int}\, h(C_i)\}_{i\in\mathbb{N}}, \{C_i'\}_{i\in\mathbb{N}}, \{r_i/2\}_{i\in\mathbb{N}})$

In the same way we can prove

<u>1.4 Proposition</u>.- $\mathrm{Diff}_W^{\Omega}(\mathbb{R}^n)$ is closed in $\mathrm{Diff}^{\Omega}(\mathbb{R}^n)$ with respect to the Whitney C^{∞}-topology.

Proof.- Let h be any element of $\mathrm{Diff}^{\Omega}(\mathbb{R}^n)$ such that $\mathrm{vol}_{\Omega}(\mathbb{R}^n - \mathrm{fix}\, h) = \infty$. There is a locally finite sequence of disjoint closed balls, $\coprod_{i\geq 1} C_i$, such that

$$\mathrm{vol}_{\Omega} \coprod_{i\geq 1} C_i = \infty, \qquad (\coprod_{i\geq 1} C_i) \cap (\coprod_{i\geq 1} h(C_i)) = \emptyset.$$

(see [1]). Let x_i, r_i be the centre and radius of the ball C_i respectively. Let C_i' be the closed ball of centre x_i and radius $r_i/2$.

The subset of $\mathrm{Diff}^{\Omega}(\mathbb{R}^n)$ given by

$$N(h;\{\mathrm{int}\, h(C_i)\}_{i\in\mathbb{N}}, \{C_i'\}_{i\in\mathbb{N}}, \{r_i/2\}) =$$

$$= \{ g \in \mathrm{Diff}^{\Omega}(\mathbb{R}^n) : g(C_i') \subset \mathrm{int}\, h(C_i) , \| D^k(h)(x) - D^k(g)(x) \| < r_i/2 \text{ for any } i \in \mathbb{N} , \text{ all } x \in C_i' , \text{ and any } k \geq 0 \}$$

is a neighbourhood of h with respect to the Whitney C^{∞}-topology that does not meet $\mathrm{Diff}_W^{\Omega}(\mathbb{R}^n)$.

§2.- Some subgroups between $\mathrm{Diff}_f^{\Omega}(\mathbb{R}^n)$ and $\mathrm{Diff}_W^{\Omega}(\mathbb{R}^n)$

Here we will define a subgroup, N, of $\mathrm{Diff}^{\Omega}(\mathbb{R}^n)$ between $\mathrm{Diff}_f^{\Omega}(\mathbb{R}^n)$ and $\mathrm{Diff}_W^{\Omega}(\mathbb{R}^n)$ that is normal in $\mathrm{Diff}_W^{\Omega}(\mathbb{R}^n)$ but not in $\mathrm{Diff}^{\Omega}(\mathbb{R}^n)$.

2.1 Definition.- Let B_i be the closed ball of $\mathbb{R}^n$ of centre the origin and radius i. We define the subset N of $\mathrm{Diff}^{\Omega}(\mathbb{R}^n)$ as follows

$$N = \{h \in \mathrm{Diff}_W^{\Omega}(\mathbb{R}^n) : \lim_{i\to\infty} \frac{\mathrm{vol}_{\Omega}((\mathrm{supp}\, h) \cap B_i)}{\mathrm{vol}_{\Omega} B_i} = 0 \}$$

2.2 Proposition.- N defined as above is a normal subgroup of $\mathrm{Diff}_W^{\Omega}(\mathbb{R}^n)$ and $\mathrm{Diff}_f^{\Omega}(\mathbb{R}^n) \subset N$.

Proof.- a) N is a group.

Let h and f be two elements of N. We have

$$\mathrm{supp}(f\circ h^{-1}) \subset (\mathrm{supp}\, f) \cup (\mathrm{supp}\, h^{-1}) = (\mathrm{supp}\, f) \cup (\mathrm{supp}\, h).$$

So,

$$\lim_{i\to\infty} \frac{\mathrm{vol}_{\Omega}((\mathrm{supp}(f\circ h^{-1}) \cap B_i)}{\mathrm{vol}_{\Omega} B_i} < \lim_{i\to\infty} \frac{\mathrm{vol}_{\Omega}((\mathrm{supp}\, f) \cap B_i)}{\mathrm{vol}_{\Omega} B_i} + $$

$$+ \lim_{i\to\infty} \frac{\mathrm{vol}_{\Omega}((\mathrm{supp}\, h) \cap B_i)}{\mathrm{vol}_{\Omega} B_i} = 0 .$$

Therefore, $f\circ h^{-1}$ lies in N.

b) N is normal in $\mathrm{Diff}^{\Omega}_{W}(\mathbb{R}^n)$.

Let h be any element of N and let g be any element of $\mathrm{Diff}^{\Omega}_{W}(\mathbb{R}^n)$. We have $\mathrm{supp}\,(g\circ h\, g^{-1}) = g(\mathrm{supp}\, h)$. Also, we have $g(\mathrm{supp}\, h) \cap B_i \subset$ $\subset ((\mathrm{supp}\, h) \cap B_i) \cup (\mathbb{R}^n - \mathrm{fix}\, g)$. Then, since $\mathrm{vol}_{\Omega}(\mathbb{R}^n - \mathrm{fix}\, g) <$ $< \infty$ we have

$$\lim_{i \to \infty} \frac{\mathrm{vol}_{\Omega}((\mathrm{supp}\, g\circ h\circ g^{-1}) \cap B_i)}{\mathrm{vol}_{\Omega} B_i} \leqslant \lim_{i \to \infty} \frac{\mathrm{vol}_{\Omega}((\mathrm{supp}\, h) \cap B_i)}{\mathrm{vol}_{\Omega} B_i} +$$

$$+ \lim_{i \to \infty} \frac{\mathrm{vol}_{\Omega}(\mathbb{R}^n - \mathrm{fix}\, g)}{\mathrm{vol}_{\Omega} B_i} = 0 .$$

Therefore, $g\circ h\circ g^{-1}$ lies in N . So, N is normal in $\mathrm{Diff}^{\Omega}_{W}(\mathbb{R}^n)$

c) $\mathrm{Diff}^{\Omega}_{f}(\mathbb{R}^n) \subset N$. It is obvious.

<u>2.3 Proposition</u>.- Let N be as in 2.1. Then, N is not normal in $\mathrm{Diff}^{\Omega}(\mathbb{R}^n)$.

<u>Proof</u>. We will construct an element h of N and we will find an element $f \in \mathrm{Diff}^{\Omega}(\mathbb{R}^n)$ such that $f\circ h\circ f^{-1}$ does not lie in N.

Let T be the standard tube of $\mathbb{R}^n$ that is

$$T = \{x \in \mathbb{R}^n \,/\, \sum_{i=2}^{n} x_i^2 \leqslant 1 \,,\, x_1 \geqslant 0\} .$$

we will construct h with support in T .

Let $\{r_i\}_{i=1}^{\infty}$ be any ordering of the positive rational numbers and let be

$$l_i = \frac{1}{i^2} ,$$

we define I_1 the following open interval of $\mathbb{R}$

$$I_1 = (r_1 - \frac{l_1}{2} , r_1 + \frac{l_1}{2})$$

and A_1 the closed subset of T

$$A_1 = \{x \in T : \sum_{i\geq 2}^{n} x_i^2 \leq 1 , r_1 - \frac{l_1}{2} \leq x_1 \leq r_1 + \frac{l_1}{2} \} .$$

Let n_2 be the smallest integer such that $r_{n_2} \notin \mathrm{cl}\, I_1$ and let $l_2' < l_{n_2}$ be a positive number such that

$$(r_{n_2} - \frac{l_2'}{2} , r_{n_2} + \frac{l_2'}{2}) \cap I_1 = \emptyset .$$

Then we call

$$A_2 = \{x \in T : \sum_{i\geq 2}^{n} x_i^2 \leq 1 , r_{n_2} - \frac{l_2'}{2} \leq x_1 \leq r_{n_2} + \frac{l_2'}{2} \}.$$

Proceeding inductively we get a family $\{A_i\}_{i \in \mathbb{N}}$ of closed subsets of T satisfying

a) $\coprod_{i\geq 1} A_i$ is dense subset of T.

b) $\mathrm{vol}_\Omega (\coprod_{i\geq 1} A_i) = \sum_{i\geq 1} \mathrm{vol}_\Omega A_i = \sum_{i \geq 1} l_i' \,\mathrm{vol}_\Omega B^{n-1} \leq$

$\leq \mathrm{vol}_\Omega B^{n-1} \sum_{i\geq 1} l_i = \mathrm{vol}_\Omega B^{n-1} \sum_{i\geq 1} \frac{1}{i^2} < \infty$

where B^{n-1} is the ball of $\mathbb{R}^{n-1}$ of centre the origin and radius 1. Thus $\bigcup_{i \geq 1} A_i$ is a subset of $\mathbb{R}^n$ with finite Ω-volume whose closure has infinite Ω-volume.

As $C = \mathbb{R} - \bigcup_{i \geq 1} I_i$ is a closed subset of $\mathbb{R}$, there is a smooth real valued function $\psi : \mathbb{R} \longrightarrow [0,\infty)$ such that $C = \psi^{-1}(0)$ (for the existence of ψ see [3]) .

Let $\phi : \mathbb{R} \longrightarrow [0,1]$ be a bump function such that $\phi(r) = 0$ for $-\infty < r \leq 0$ or $1 \leq r < +\infty$. We define, for any $x = (x_1,\dots,x_n) \in \mathbb{R}^n$ the matrix

$$M(x) = \begin{pmatrix} I & & 0 & \\ & & & \\ & \cos(\phi(\sum_{i \geq 2}^{n} x_i^2)\ \psi(x_1) & & -\sin\phi(\sum_{i \geq 2}^{n} x_i^2)\ \psi(x_1) \\ 0 & & & \\ & \sin(\phi(\sum_{i \geq 2}^{n} x_i^2)\ \psi(x_1) & & \cos\phi(\sum_{i \geq 2}^{n} x_i^2)\ \psi(x_1) \end{pmatrix}$$

Now, we define $h : \mathbb{R}^n \longrightarrow \mathbb{R}^n$ by $h(x) = x \cdot M(x)$. Clearly, it is a volume preserving diffeomorphism such that $\mathbb{R}^n - \text{fix} = \coprod_{i \geq 1} A_i$ and $\text{supp } h = T$. Therefore h lies in $\text{Diff}_W^\Omega(\mathbb{R}^n)$.

Furthermore, h is an element of N since

$$\lim_{i \to \infty} \frac{\text{vol}_\Omega((\text{supp } h) \cap B_i}{\text{vol}_\Omega B_i} = \lim_{i \to \infty} \frac{\text{vol}_\Omega(T \cap B_i)}{\text{vol}_\Omega B_i} \leq$$

$$\leq \lim_{i \to \infty} \frac{(\text{vol}_\Omega B^{n-1}) \cdot i}{\text{vol}_\Omega B_i} = \lim_{i \to \infty} \frac{i(\text{vol}_\Omega B^{n-1})}{i^n(\text{vol}_\Omega B^n)} =$$

$$= \lim_{i \to \infty} \frac{\text{vol}_\Omega B^{n-1}}{i^{n-1} \text{vol}_\Omega B^n} = 0$$

where B^n is the closed ball of $\mathbb{R}^n$ of centre the origin and radius 1.

Now we will find an element f of $\text{Diff}^\Omega(\mathbb{R}^n)$ such that $f \circ h \circ f^{-1}$ does not lie in N .

Let V be the following subset of $\mathbb{R}^n$

$$V = \{x \in \mathbb{R}^n : x_1 \geq 0\} .$$

There is an element f of $\text{Diff}^\Omega(\mathbb{R}^n)$ such that $f(T) = V$ (see [1]). Therefore

$$\lim_{i \to \infty} \frac{\text{vol}_\Omega((\text{supp } f \circ h \circ f^{-1}) \cap B_i}{\text{vol}_\Omega B_i} = \lim_{i \to \infty} \frac{\text{vol}_\Omega(f(\text{supp } h) \cap B_i}{\text{vol}_\Omega B_i} =$$

$$= \lim_{i \to \infty} \frac{\text{vol}_\Omega(V \cap B_i)}{\text{vol}_\Omega B_i} = \lim_{i \to \infty} \frac{1/2 \text{ vol}_\Omega B_i}{\text{vol}_\Omega B_i} = \frac{1}{2} .$$

So, $f \circ h \circ f^{-1}$ is not an element of N .

2.4 Note.- It is clear that if in the definition of N (2.1) we put some growth condition we get a family of normal subgroups of $\mathrm{Diff}^{\Omega}_{W}(\mathbb{R}^n)$ between $\mathrm{Diff}^{\Omega}_{f}(\mathbb{R}^n)$ and $\mathrm{Diff}^{\Omega}_{W}(\mathbb{R}^n)$ no normals in $\mathrm{Diff}^{\Omega}_{W}(\mathbb{R}^n)$.

REFERENCES

[1] F. Mascaró "Normal subgroups of $\mathrm{Diff}^{\Omega}(\mathbb{R}^n)$", to appear in Trans. A.M.S.

[2] F. Mascaró "Normal subgroups of $\mathrm{Diff}^{\Omega}(\mathbb{R}^3)$", preprint.

[3] R. Narasimhan R. "Analisis on real and complex manifolds" Advanced Studies in Pure Math. (1968), North-Holland.

[4] W. Thurston "On the structure of the group of volume preserving diffeomorphisms", to appear.

SOME INTEGRAL INVARIANTS OF PLANE FIELDS ON RIEMANNIAN MANIFOLDS

A. Montesinos
Departamento de Geometría y Topología
Facultad de Matemáticas
Burjasot (Valencia). Spain

1. INTRODUCTION

Let (M,g,P) be an orientable riemannian almost product manifold, i.e. $P^2 = 1$, $g(PX,PY) = g(X,Y)$. We put $TM = \mathcal{V} \oplus \mathcal{H}$, being $\mathcal{V}$ and $\mathcal{H}$ the subbundles corresponding to the eigenvalues 1 and -1 of P, respectively, and $p = \operatorname{rank} \mathcal{V}$, $q = \operatorname{rank} \mathcal{H}$, $n = p + q$. The projectors upon $\mathcal{V}$ and $\mathcal{H}$ are denoted by v and h, and the volume form by ω.

If ∇ is the Levi-Civita connection, the connection given by

$$\tilde{\nabla}_X Y = \nabla_X Y + \frac{1}{2} P(\nabla_X P) Y$$

satisfies $\tilde{\nabla} g = 0$, $\tilde{\nabla} P = 0$. Hence, it induces connections on the subbundles $\mathcal{V}$ and $\mathcal{H}$. If $\mathcal{V}$ (or $\mathcal{H}$) happens to be a foliation, then the connection induced by $\tilde{\nabla}$ on the leaves is the Levi-Civita connection on them.

F. Brito, R. Langevin and H. Rosenberg [2] have proved that if $q = 1$, $\mathcal{V}$ is a transversaly orientable foliation and M is of constant sectional curvature, then the integral over M (closed) of the mean r-dimensional Lipschitz-Killing curvature of $\mathcal{V}$ does not depend on $\mathcal{V}$. Also, F. Brito [1] proved in the same conditions upon $\mathcal{V}$, that the scalar curvature of $\mathcal{V}$ does not depend on $\mathcal{V}$ if M is a closed Einstein manifold. We shall prove the same results for a plane field $\mathcal{V}$ of codimension 1, orientable or not, and also the following

THEOREM 1: Let M be closed and of constant sectional curvature c. Let K_v and K_h be the scalar curvatures of $(\tilde{\nabla}, \mathcal{V})$ and $(\tilde{\nabla}, \mathcal{H})$. Then

$$\int_M (K_v + K_h)\,\omega = c\left(2\binom{n}{2} - pq\right) \operatorname{Vol} M .$$

2. DOUBLE FIELDS

$\{e_i\} = \{e_a,e_u\}$, $e_a \in V$, $e_u \in H$, will denote an adapted oriented orthonormal local frame and $\{\theta_i\} = \{\theta_a,\theta_u\}$ its dual. Thus, the volume form is $\omega = \theta_1\wedge\ldots\wedge\theta_n$.

Let $\mathcal{D}^{r,s}$ denote the module of tensor fields of type $(0,r+s)$ that are skewsymmetric in their first r arguments. In $\mathcal{D} = \oplus\, \mathcal{D}^{r,s}$ we define a multiplication "$\cdot$" which is the exterior product in the first arguments and the tensor product in the last ones. $\hat{\mathcal{D}} = \oplus\, \hat{\mathcal{D}}^{r,s}$ denotes the algebra of double forms [3], [4], [5], with product "$\wedge$".

Let α , $\beta \in \mathcal{D}$ (or $\hat{\mathcal{D}}$ in the second case). We put

$$\alpha \underset{k}{\cdot} \beta = \alpha(\ ;e_{i_1},\ldots,e_{i_k},\ )\cdot\beta(\ ;e_{i_1},\ldots,e_{i_k},\ )$$

$$\alpha \underset{k}{\wedge} \beta = \alpha(\ ;e_{i_1},\ldots,e_{i_k},\ )\wedge\beta(\ ;e_{i_1},\ldots,e_{i_k},\ )\ ,$$

or zero if α and/or β belong to $\mathcal{D}^{r,s}$ with $s < k$.

If $\alpha \in \mathcal{D}^{r,s}$ (note that $\hat{\mathcal{D}}^{r,s} \subset \mathcal{D}^{r,s}$) and ∇ is a linear connection on M, the covariant differential of α , $D\alpha \in \mathcal{D}^{r+1,s}$, is defined by

$$(D\alpha)(X_o,\ldots,X_r;\ ) = \sum_{i=o}^{r} (-1)^i\, \nabla_{X_i}\alpha(X_o,\ldots,\hat{X}_i,\ldots,X_r;\ ) + \sum_{i<j} (-1)^{i+j}\, \alpha([X_i,X_j],X_o,\ldots,\hat{X}_i,\ldots,\hat{X}_j,\ldots,X_r;\ )\ .$$

Let ∇ be such that $\nabla g = 0$. We define $R \in \hat{\mathcal{D}}^{2,2}$ as :

$$R(X,Y;Z,W) = g(\nabla_X\nabla_Y Z - \nabla_Y\nabla_X Z - \nabla_{[X,Y]} Z\ ,\ W)$$

The following results are easily verified.

2.1 PROPOSITION: Let $\alpha \in \mathcal{D}^{r,s}$, $\beta \in \mathcal{D}$, $\nabla g = 0$. Then

i) $D(\alpha \underset{k}{\cdot} \beta) = D\alpha \underset{k}{\cdot} \beta + (-1)^r\, \alpha \underset{k}{\cdot} D\beta$, and the same in $(\hat{\mathcal{D}},\wedge)$;

ii) if $\alpha \in \hat{\mathcal{D}}$, then $D^2\alpha = R \underset{1}{\wedge} \alpha$.

3. THE CHARACTERISTIC CONNECTION

The connection $\tilde{\nabla}$ given in §1 is called sometimes the characteristic connection of the riemannian almost product structure (M,g,P). As it is well known, $\tilde{\nabla}g = 0$ and $\tilde{\nabla}P = 0$. Hence $\tilde{R}(\ , \ ;A,X) = 0$ if $A \in V$ and $X \in H$. If A,B are vector fields in V and V is a foliation, we find immediately

$$\tilde{\nabla}_A B - \tilde{\nabla}_B A - [A,B] = 0.$$

Hence, $\tilde{\nabla}$ is the Levi-Civita connection in the leaves of V. This justifies to take $\tilde{\nabla}$ for defining the Lipschitz-Killing curvatures of the plane fields V and H although they were not foliations.

We define $A \in \mathcal{D}^{0,2}$, $B \in \hat{\mathcal{D}}^{1,1}$, $W \in \hat{\mathcal{D}}^{1,2}$ by:

$$A(X,Y) = g(PX,Y)$$
$$B(X;Y) = A(X,Y)$$
$$W(X;Y,Z) = g(P(\nabla_X P)Y,Z).$$

A direct calculation gives

$$\tilde{R} = R + \frac{1}{2}DW + \frac{1}{8}\, W \overset{\wedge}{\underset{1}{}} W$$

We have the following properties:

3.1. PROPOSITION:

i) $W(X;Y,Z) = - W(X;PY,PZ)$; hence $W(X;e_a,e_b) = W(X;e_u,e_v) = 0$;

ii) $W(X;Y,PZ) = DA(X;Y,Z)$;

iii) $W \overset{\wedge}{\underset{1}{}} W = 2(DA) \overset{\cdot}{\underset{1}{}} (DA) = 2D(A \overset{\cdot}{\underset{1}{}} DA) - 2A \overset{\cdot}{\underset{1}{}} D^2A$;

iv) $(D^2A)(\ ;X,Y) = 2R(\ ;vX,hY) - 2R(\ ;hX,vY)$

v) $A \overset{\cdot}{\underset{1}{}} DA = -W$

vi) $(A \overset{\cdot}{\underset{1}{}} D^2A)(\ ;X,Y) = 2R(\ ;vX,hY) + 2R(\ ;hX,vY)$.

Proof.- Simple computation. For example

$$D^2A(X,Y;Z,S) = (\nabla_X DA(Y; \ ) - \nabla_Y DA(X; \ ) - DA([X,Y] \ ; \ ))(Z,S) =$$
$$= (\nabla_X g(\nabla_Y P, \ ) - \nabla_Y g(\nabla_X P, \ ) - g(\nabla_{[X,Y]} P, \ ))(Z,S) =$$

$= g(R(X,Y)P,\ )(Z,S) =$

$= R(X,Y;PZ,S) - R(X,Y;Z,PS) =$

$= 2R(X,Y;vZ,hS) - 2R(X,Y;hZ,vS).$

3.2. COROLLARY: $\tilde{R} = R + \frac{1}{4} DW - \frac{1}{2} R(\ ;v\ ,h\ ) - \frac{1}{2} R(\ ;h\ ,v\ ).$

For each integer m, $0 \leq m \leq n$, we define $\omega^{(m)} \in \hat{\mathcal{D}}^{n-m,m}$ by

$$\omega^{(m)}(X_1,\ldots,X_{n-m};X_{n-m+1},\ldots,X_n) = \omega(X_1,\ldots,X_n),$$

and clearly we have $D\omega^{(m)} = 0$.

The following formulae are purely combinatorial.

3.3 PROPOSITION: i) $\omega^{(2r)} \underset{2r}{\wedge} g^r_{vh} = (2r)!(r!)^2 \binom{p}{r}\binom{q}{r}\omega$, where

$$g_{vh} = \frac{1}{4}(g^2 - B^2),\ g^r = g\wedge\ldots\wedge g \quad (r \text{ times}), \text{ etc.}$$

ii) if $\alpha \in \hat{\mathcal{D}}^{r,m-s}$, then

$$\omega^{(m)} \underset{m}{\wedge} (g^s \wedge \alpha) = (s!)^2 \binom{m}{s}\binom{n-m+s}{s} \omega^{(m-s)} \underset{m-s}{\wedge} \alpha.$$

iii) if $\alpha \in \hat{\mathcal{D}}^{\cdot\cdot,r}$, $\beta \in \hat{\mathcal{D}}^{\cdot\cdot,s}$, then

$$\omega^{(r+s)} \underset{r+s}{\wedge} (\alpha\wedge\beta) = \binom{r+s}{r} (\omega^{(r+s)} \underset{r}{\wedge} \alpha) \underset{s}{\wedge} \beta.$$

iv) $\omega^{(r)} \underset{r}{\wedge} B^r = (r!)^2 \sum_{i=o}^{r} (-1)^i \binom{p}{r-i}\binom{q}{i}\omega.$

4. THE MAIN RESULTS

We compute $\omega^{(2)} \underset{2}{\wedge} \tilde{R}$.

$$\omega^{(2)} \underset{2}{\wedge} \tilde{R} = \omega(\ ;e_i,e_j) \wedge \tilde{R}(\ ;e_i,e_j) =$$

$$= \tilde{R}(e_i,e_j;e_i,e_j)\omega =$$

$$= (\tilde{R}(e_a,e_b;e_a,e_b) + \tilde{R}(e_u,e_v;e_u,e_v))\omega =$$

$$= -(K_v + K_h)\omega,$$

where K_v (resp. K_h) is the scalar curvature of the connection $\tilde{\nabla}$ along the plane field V (resp. H).

THEOREM 1: Let M be closed and of constant sectional curvature c. Then

$$\int_M (K_v + K_h) = c\left\{2\binom{n}{2} - pq\right\} \mathrm{Vol}\, M .$$

Proof.- We have

$$(K_v + K_h) = -\omega^{(2)} \underset{2}{\wedge} (R + \frac{1}{4} DW - \frac{1}{2} R(\ ;v \ ,h \ ) - \frac{1}{2} R(\ ;h \ ,v \ )) .$$

Since (M,g) is of constant sectional curvature c, we can put $R = - \frac{1}{2} cg^2$.

Thus $\frac{1}{2} R(\ ;v \ ,h \ ) + \frac{1}{2} R(\ ;h \ ,v \ ) = - \frac{1}{2} cg_{vh}$, and

$$- \frac{1}{2} c\omega^{(2)} \underset{2}{\wedge} g_{vh} = -pqc\omega .$$

Also

$$-\omega^{(2)} \underset{2}{\wedge} R = \frac{1}{2} c\omega^{(2)} \underset{2}{\wedge} g^2 = 2\binom{n}{2} c\omega .$$

Therefore

$$(K_v + K_h)\omega = 2\binom{n}{2} c\omega - pqc\omega - \frac{1}{4} \omega^{(2)} \underset{2}{\wedge} DW =$$

$$= c\left\{2\binom{n}{2} - pq\right\}\omega - \frac{1}{4} (-1)^n d(\omega^{(2)} \underset{2}{\wedge} W) .$$

Hence, if M is closed (compact and without boundary), we have our claim by Stokes' Theorem.

THEOREM 2: Let (M,g) be a closed Einstein manifold, i.e. Ricci $= \frac{c}{n} g$.

If $q = 1$, then $\int_M K_v \omega = \frac{c(n-1)}{n} \mathrm{Vol}\, M$.

Proof.- if $q = 1$, then $K_h = 0$, and as before

$$K_v\omega = -\frac{1}{4}(-1)^n d(\omega^{(2)} \underset{2}{\wedge} W) - \omega^{(2)} \underset{2}{\wedge} R + \frac{1}{2}\omega^{(2)} \underset{2}{\wedge}(R(\ ;v \ ,h \ ) + $$

$$+ R(\ ;h \ ,v \ )) .$$

But $-\omega^{(2)} \underset{2}{\wedge} R = c\omega$, and

$$\frac{1}{2}\omega^{(2)} \underset{2}{\wedge}(R(\ ;v \ ,h \ ) + R(\ ;h \ ,v \ )) =$$

$$= R(e_a,e_u;e_a,e_u)\,\omega = R(e_i,e_u;e_i,e_u)\,\omega = -\operatorname{Ricci}(e_u,e_u)\,\omega =$$

$$= -\frac{c}{n}\omega .$$

Therefore

$$K_v = \frac{c(n-1)}{n}\omega - \frac{1}{4}(-1)^n d(\omega^{(2)} \underset{2}{\wedge} W),$$

and our claim follows

THEOREM 3: Let M be closed and of constant sectional curvature. Let γ_r be the mean r-dimensional Lipschitz-Killing curvature of V and $q = 1$. Then

$$\int_M \gamma_r\omega = c^r \sum_{k=0}^{r} \frac{\binom{r}{k}\left(\binom{n-1}{2k}\right)}{\binom{n-1}{2k}} \operatorname{Vol} M .$$

Proof.- As in Theorem 1, we have $\tilde{R} = -\frac{1}{2}cg^2 + \frac{1}{2}cg_{vh} + \frac{1}{4}DW$. Now, $DW = 2cg_{vh} - 2(D\theta_n)^2$, as it is easily verified. Thus

$$\tilde{R} = -\frac{1}{2}(cg_v^2 + (D\theta_n)^2) ,$$

where $g_v(X;Y) = g(vX,vY)$. Therefore

$$\Phi^{(r)}(\tilde{R}) = \omega^{(2r)} \underset{2r}{\wedge} \tilde{R}^r =$$

$$= (-\frac{1}{2})^r \sum_{k=0}^{r} c^{r-k}\binom{r}{k}\omega^{(2r)} \underset{2r}{\wedge}(g_v^{2r-2k} \wedge (D\theta_n)^{2k}) .$$

Since $(D\theta_n)(\ ;e_n) = 0$, we have:

$$\omega^{(r+s)} \underset{r+s}{\wedge} (g_v^r \wedge (D\theta_n)^s) = (r!)^2 \binom{r+s}{r} \binom{n-s-1}{r} \omega^{(s)} \underset{s}{\wedge} (D\theta_n)^s .$$

Therefore

$$\Phi^{(r)}(\tilde{R}) = (-\tfrac{1}{2})^r \sum_{k=0}^{r} c^{r-k} \binom{r}{k} \frac{(2r)!(n-1-2k)!}{(2k)!(n-1-2r)!} \omega^{(2k)} \underset{2k}{\wedge} (D\theta_n)^{2k} .$$

Having in mind that $\theta_n \wedge D^2\theta_n = -cg_{vh}$, we obtain:

$$\omega^{(2k)} \underset{2k}{\wedge} (D\theta_n)^{2k} = (-1)^n\, d\Big(\omega^{(2k)} \underset{2k}{\wedge} (\theta_n \wedge (D\theta_n)^{2k-1})\Big) +$$

$$+ c(2k)(2k-1)^2(n-2k+1)\omega^{(2k-2)} \underset{2k-2}{\wedge} (D\theta_n)^{2k-2} .$$

The form $\omega^{(2k)} \underset{2k}{\wedge} (\theta_n \wedge (D\theta_n)^{2k-1})$ does not depend on the choice between $\pm\theta_n$; thus, it is a global form, and in cohomology we have

$$\omega^{(2k)} \underset{2k}{\wedge} (D\theta_n)^{2k} = ((2k)!)^2 \left(\!\binom{n-1}{2k}\!\right) c^k \omega ,$$

where

$$\left(\!\binom{n-1}{2k}\!\right) = \frac{(n-1)(n-3)\dots(n-2k+1)}{2k(2k-2)\dots4\cdot2} ,$$

or 1 if $k = 0$. Therefore in cohomology

$$\Phi^{(r)}(\tilde{R}) = (-\tfrac{1}{2})^r (2r)!c^r \sum_{k=0}^{r} \binom{r}{k} \left(\!\binom{n-1}{2k}\!\right) \frac{(n-1-2k)!(2k)!}{(n-1-2r)!} \omega .$$

Since (cfr. [2]):

$$\gamma_r \omega = \frac{(-2)^r}{((2r)!)^2 \binom{n-1}{2r}} \Phi^{(r)}(\tilde{R}) ,$$

where γ_r is the r-dimensional Lipschitz-Killing curvature of V, the theorem follows by substitution.

Remark.- The use of this technique does not force to take V transversally orientable as in [2]. If $c \neq 0$, n-1 must be even due to the existence of H; then, the above formula coincides with [2], 2.3.3. Or if $n = 2$ and c is not constant, the last step in the proofs leads to $\chi(M) = 0$.

REFERENCES

[1] F. Brito, Une obstruction géométrique à l'existence de feuilletages de codimension 1 totalement géodésiques, J. Differential Geometry 16 (1981), 675-684.

[2] F. Brito, R. Langevin, H. Rosenberg, Intégrales de courbure sur des variétés feuilletées, J. Differential Geometry 16 (1981), 19-50.

[3] A. Gray, Some relations between curvature and characteristic classes, Math. Ann. 184 (1970), 257-267.

[4] R.S. Kulkarni, On the Bianchi identities, Math. Ann. 199 (1972), 175-204.

[5] G. de Rham, On the area of complex manifolds, Seminar on Several Complex Variables, Institute for Advanced Study, 1957.

A SCHUR-LIKE LEMMA FOR THE NK-MANIFOLDS OF CONSTANT TYPE

A. M. Naveira
Departamento de Geometría y Topología
Facultad de Matemáticas
Valencia. SPAIN

In this note we find a Schur-like theorem for Nearly-Kähler manifolds (NK-manifolds) and solve a question posed by A. Gray in [1].

After getting this result we have known that Kirichenko in [5] has solved the same question, but we feel that our proof can be useful for anyone working on almost Hermitian manifolds.

We denote by J the tensor defining the almost complex structure and by ∇ the Riemannian connection of g, the metric tensor of M, which is compatible with J, in the sense that $g(JX,JY) = g(X,Y)$. In this case, M is called an almost Hermitian manifold, and M is said to be an NK-manifold (or Tachibana manifold) if it verifies

$$\nabla_X(J)X = 0$$

for all $X \in \chi(M)$, where $\chi(M)$ denotes the Lie algebra of vector fields on M.

All geometric objects that we consider in this note will be supposed C^∞.

Proposition 1. [1], [4].- The Riemannian curvature operator R of an NK- manifold verifies

i) $R_{WXYZ} - R_{WXJYJZ} = g(\nabla_W(J)X, \nabla_Y(J)Z)$

ii) $R_{WXYZ} - R_{WXJYJZ} = R_{WJXYJZ} + R_{WJXJYZ}$

iii) $R_{WXYZ} = R_{JWJXJYJZ}$

for all $W, X, Y, Z \in \chi(M)$.

Let M be an almost Hermitian manifold. Then M is said to be of

<u>constant type</u> at $m \in M$ provided that for all $x \in M_m$, we have

$$\| \nabla_x(J)y \| = \| \nabla_x(J)z \|$$

whenever $g(x,y) = g(Jx,y) = g(x,z) = g(Jx,z) = 0$ and $\|y\| = \|z\|$. If this holds for all $m \in M$ we say that M is <u>pointwise constant type</u>. Finally, if M has pointwise constant type and for $X, Y \in \chi(M)$ with $g(X,Y) = g(JX,Y) = 0$ the function $\| \nabla_X(J)Y \|$ is constant whenever $\|X\| = \|Y\| = 1$ then we say that M has <u>global constant type</u>.

The following Proposition is useful for our study.

<u>Proposition 2</u>. [1].- <u>Let M be an NK-manifold. Then M has pointwise constant type if and only if there exists a function α such that</u>

$$\| \nabla_W(J)X \|^2 = \alpha\{ \|W\|^2 \|X\|^2 - \langle W,X\rangle^2 - \langle W,JX\rangle^2 \} \qquad (1)$$

<u>for all $W, X \in \chi(M)$. Furthermore, M has global constant type if and only if (1) holds with a constant function α</u>.

Since every NK-manifold of dimension four is a Kahler manifold, [2], we can consider always $\dim_{\mathbb{R}} M \geq 6$.

The proof of the following Lemma is straightforward by linearization in (1).

<u>Lemma 3</u>.- <u>Let M be an NK-manifold with pointwise constant type, then</u>

$$\begin{aligned} & g(\nabla_W(J)Z, \nabla_X(J)Y) + g(\nabla_X(J)Z, \nabla_W(J)Y) = \\ & = R_{WZXY} + R_{XZWY} - R_{WZJXJY} - R_{XZJWJY} = \\ & = \alpha\{2g(W,X)g(Z,Y) - g(W,Z)g(X,Y) - g(X,Z)g(W,Y) - \\ & - g(W,JZ)g(X,JY) - g(X,JZ)g(W,JY)\} \end{aligned} \qquad (2)$$

Now, taking the covariant derivative in (2) we have

<u>Proposition 4</u>.- <u>Let M be an NK-manifold of pointwise constant type, then</u>

$$\nabla_U(R)_{WZXY} + \nabla_U(R)_{XZWY} - \nabla_U(R)_{WZJXJY} - \nabla_U(R)_{XZJWJY} -$$

$$- R_{WZ\nabla_U(J)X} - R_{WZJX\nabla_U(J)Y} - R_{XZ\nabla_U(J)WJY} - R_{XZJW\nabla_U(J)Y} =$$

$$\begin{aligned} &= (U\alpha)\{2g(W,X)g(Z,Y) - g(W,Z)g(X,Y) - g(X,Z)g(W,Y) - \\ &- g(W,JZ)g(X,JY) - g(X,JZ)g(W,JY)\} + \\ &- \alpha\{ g(W,\nabla_U(J)Z)g(X,JY) + g(W,JZ)g(X,\nabla_U(J)Y) + \\ &+ g(X,\nabla_U(J)Z)g(W,JY) + g(X,JZ)g(W,\nabla_U(J)Y)\} \end{aligned} \tag{3}$$

<u>Proposition 5</u>.- <u>Let M be an NK-manifold of pointwise constant type, then</u> :

$$\begin{aligned} &\nabla_U(R)_{XZWY} - \nabla_U(R)_{XZJWJY} + \nabla_U(R)_{JXZWJY} + \nabla_U(R)_{JXZJWY} - \\ &- g(\nabla_W(J)Z, \nabla_{\nabla_U(J)X}(J)JY) - g(\nabla_W(J)Z, \nabla_{JX}(J)\nabla_U(J)Y) - \\ &-R_{XZ\nabla_U(J)WJY} - R_{XZJW\nabla_U(J)Y} + R_{JXZ\nabla_U(J)WY} - R_{JXZJW\nabla_U(J)JY} = \\ &= 2(U\alpha)\{g(W,X)g(Z,Y) - g(W,Z)g(X,Y) - \\ &- g(W,JZ)g(X,JY) + g(W,JX)g(Z,JY)\} - \\ &- \alpha\{2g(W,\nabla_U(J)Z)g(X,JY) + g(X,\nabla_U(J)Z)g(W,JY) + \\ &+ g(X,JZ)g(W,\nabla_U(J)Y) - g(JX,\nabla_U(J)Z)g(W,Y) + \\ &+ g(X,Z)g(W,J\nabla_U(J)Y)\} . \end{aligned} \tag{4}$$

In order to obtain (4), we change in (3) X and Y by JX and JY respectively and we add the new expression to (3).

<u>Lemma 6</u>.- <u>If M is an NK-manifold, then</u>

$$\begin{aligned} &\nabla_U(R)_{XZWY} - \nabla_U(R)_{XZJWJY} - \nabla_U(R)_{JXZJWY} - \nabla_U(R)_{JXZWJY} - \\ &- R_{XZ\nabla_U(J)WJY} - R_{XZJW\nabla_U(J)Y} - R_{\nabla_U(J)XZJWY} - \\ &- R_{JXZ\nabla_U(J)WY} - R_{\nabla_U(J)XZWY} - R_{JXZW\nabla_U(J)Y} = 0 . \end{aligned} \tag{5}$$

In this case, the proof follows by taking the covariant derivative in Prop. 1.i).

<u>Lemma 7</u>.- <u>Let M be an NK-manifold of pointwise constant type, then</u>

$$\begin{aligned} &2\nabla_U(R)_{XZWY} - 2\nabla_U(R)_{XZJWJY} = \\ &= 2(U\alpha)\{g(W,X)g(Z,Y) - g(W,Z)g(X,Y) - \\ &- g(W,JZ)g(X,JY) + g(W,JX)g(Z,JY)\} - \end{aligned}$$

$$\begin{aligned}
&-\alpha\{2g(W,\nabla_U(J)Z)g(X,JY) + g(X,\nabla_U(J)Z)g(W,JY) + \\
&+ g(X,JZ)g(W,\nabla_U(J)Y) - g(JX,\nabla_U(J)Z)g(W,Y) + \\
&+ g(X,Z)g(W,J\nabla_U(J)Y)\} + 2R_{XZ\nabla_U(J)WJY} + 2R_{XZJW\nabla_U(J)Y} + \\
&+ g(\nabla_W(J)Z,\nabla_{\nabla_U(J)X}(J)JY) + g(\nabla_W(J)Z,\nabla_{JX}(J)\nabla_U(J)Y) + \\
&+ g(\nabla_{\nabla_U(J)X}(J)Z,\nabla_{JW}(J)Y) + g(\nabla_{JX}(J)Z,\nabla_W(J)\nabla_U(J)Y)
\end{aligned} \tag{6}$$

The proof follows from (4) and (5).

Now, we are ready to prove our main result and therefore to solve the question stated by A. Gray, [1], about the existence of a Schur-like lemma on the pointwise constant type.

Theorem 8.- Let M be an NK-manifold of pointwise constant type, then M is also of global constant type.

Proof. We can use a similar method to the one used in the proof of Schur's Lemma for the NK-manifolds, [6]. That is, taking the cyclic sommation in (6) over U, X and Z, using the second Bianchi identity, the properties of ∇J and taking X,JX,Y and JY mutually orthonormal with $U = JX$, $W = X$, $Z = Y$, we have

$$\begin{aligned}
&(X\alpha) + R_{XYJX\nabla_{JX}(J)Y} + R_{JXX\nabla_Y(J)XJY} + R_{YJXJX\nabla_X(J)Y} = \\
&=(X\alpha) - R_{XYJXJ\nabla_X(J)Y} + R_{JXXJY\nabla_X(J)Y} + R_{YJXJX\nabla_X(J)Y} = \\
&=(X\alpha) - R_{XYJXJ\nabla_X(J)Y} - R_{JXXYJ\nabla_X(J)Y} - R_{YJXXJ\nabla_X(J)Y} = \\
&=(X\alpha) = 0 .
\end{aligned}$$

REFERENCES

[1] A.Gray. Nearly Kähler manifolds.J. Diff. Geometry 4, 283-309, (1970).

[2] A. Gray. Almost complex submanifolds of the six sphere. Proc. Amer. Math. Soc. 20, 277-279, (1969).

[3] A. Gray. The structure of Nearly Kähler manifolds. Math. Ann. 223, 233-248, (1976).

[4] A. Gray. Curvature identities for Hermitian and Almost Hermitian manifolds. Tôhoku Math. J. 28, 601-612, (1976).

[5] V.F. Kirichenko. K-spaces of constant type. (russian) Sibirski Matematicheski Journal XVII, (1976).

[6] A.M. Naveira - L.M. Hervella. Schur's theorem for Nearly Kähler manifolds". Proc. Amer. Math. Soc. 49, 421-425, (1975).

[7] S. Kobayashi - K. Nomizu. Foundations of differential geometry. 2 vol. Interscience. New York, 1963, 1969.

COMPACT HAUSDORFF FOLIATIONS

M. Nicolau and A. Reventós
Secció de Matemàtiques
Universitat Autònoma Barcelona (SPAIN)

§0. Introduction

In [4] Haefliger defines for each oriented foliation F a linear operartor $\int_F$ (the integration along the leaves) with similar properties to those of the integration over the fibres on fibre bundles.

We use this operator to obtain a differentiable version of the Gysin sequence and Euler class for a compact Hausdorff sphere foliation. From this we show that the vanishing of this class is related to the existence of a Riemannian transverse foliaton.

We also use the integration along the leaves to obtain an integral formula, of Gauss-Bonnet tipe, which generalizes a result by Duchamp contained is his thesis [1].

§1. The integration along the leaves

In this paper M will denote an oriented compact connected smooth manifold of dimension $p + q$. Let F be a smooth oriented Hausdorff compact foliation on M, of codimension q, i.e. all leaves of F are compact, smoothly oriented and the leaf space $B = M/F$ is Hausdorff.

In this situation (cf. for instance [2]) there is a generic leaf L and an open dense subset of M where the leaves are all difeomorphic to L. Moreover, for each leaf L_o there is a finite subgroup G of $0(q)$ acting freely on L and a diffeomorphism $\phi : L \times_G D \longrightarrow V$, where V is an open neighborhood of L_o and D is the unit ball of $\mathbb{R}^q$, which preserves leaves if one takes the foliation on $L \times_G D$ whose leaves are the quotient of the submanifolds $L \times \{\text{points}\}$.

It follows from this that B is, in a natural way a V-manifold and that the canonical projection $\pi : M \longrightarrow B$ defines a V-fibre space structure. For definitions and general properties of V-manifolds we refer to [5].

The above local description of F enable us to interpret the integration along the leaves introduced by Haefliger in [4] as a linear map of degree $-p$,

$$\oint_F : A^*(M) \longrightarrow A^*(B)$$

where $A^*(B)$ denotes the algebra of V-forms on the V-manifold B. In fact, if ω is a k-form on M we can compute $\oint_F \omega$ in the following way : in each local model $L \times_G D$ the k-form $\phi^*\omega$ can be regarded as a G-invariant k-form on $L \times D$. The integration of this form along the fibres in the trivial bundle $L \times D \longrightarrow D$ gives us a G-invariant $(k - p)$ - form on D. Now the canonical structure of V-fibre space of $\pi : M \longrightarrow B$ tells us that the above construction defines a V-form on the V-manifold B. This V-form is $\oint_F \omega$.

We list here some properties of this operator we shall use later.

Proposition 1. *(a) The integration along the leaves commutes with the exterior derivative*

(b) Let τ be a r-V-form on B and ω a s-form on M. Then $\oint_F(\pi^*\tau \wedge \omega) = \tau \wedge \oint_F \omega$.

(c) The V-manifold B is orientable and for any n-form ω on M, we have $\int_M \omega = \int_B \oint_F \omega$.

All the above statements follow directly from the analogous properties of the integration over the fibres on fibre bundles.

§2. The Gysin sequence

In this paragraph F will be an oriented sphere foliation, i.e. an oriented Hausdorff compact foliation with generic leaf the sphere of dimension p. We also assume M compact. This situation includes Seifert manifolds.

Since F is oriented, there is a p-form ω on M which induces a volume element on each leaf. As $\oint_F \omega \neq 0$, we can assume $\oint_F \omega = 1$. Thus, for each $\tau \in A^*(B)$, we have $\oint_F (\pi^*\tau \wedge \omega) = \tau$, and $\oint_F : A^r(M) \longrightarrow A^{r-p}(B)$ is exhaustive. Let K^r denote the kernel of $\oint_F$. It follows from Proposition 1 (a) that $dK^r \subset K^{r+1}$. Thus we obtain the exact sequence of differential spaces

$$(1) \qquad 0 \longrightarrow K^* \longrightarrow A^*(M) \xrightarrow{f_F} A^*(B) \longrightarrow 0$$

We denote by ∂ the connecting homomorphism associated to this sequence. We also have Im $\pi^* \subset K^*$ and the commutative diagram

$$\begin{array}{ccc} A^r(B) & \xrightarrow{\pi^*} & K^r \\ d\downarrow & & \downarrow d \\ A^{r+1}(B) & \xrightarrow{\pi^*} & K^{r+1} \end{array}$$

induces, for each r, a morphism $\pi^\#_r : H^r(B) \longrightarrow H^r(K^*)$. $H^r(B)$ denotes the de Rham cohomology of the V-manifold B.

Proposition 2. $\pi^\#: H^*(B) \longrightarrow H^*(K^*)$ *is an isomorphism.*

Proof. Let $\phi : S^p \times_G D \longrightarrow V$ be a local model and let W be a saturated open subset of V. For each $\beta \in A(S^p)$ set $\beta_G = \frac{1}{|G|} \sum_{\sigma \in G} \sigma^*\beta$, where $|G|$ denotes the order of G. Then $\beta_G \times 1$ is a G-invariant form on $S^p \times D$ and thus it can be regarded as a form on $S^p \times_G D$.

Define $k : A^*(\pi W) \otimes A^*(S^p) \longrightarrow A^*(W)$ to be the linear map such that $k(\alpha \otimes \beta) = \pi^*\alpha \wedge (\gamma | W)$, where γ is the form on V corresponding to $\beta_G \times 1$. Since k is a morphism of graded differential algebras, it induces a morphism $k^\# : H^*(A^*(\pi W) \otimes A^*(S^p)) \longrightarrow H^*(W)$.

As in the case of sphere bundles the proposition reduces to prove that $k^\#$ is an isomorphism (see for instance [3]). Note that as W is not a product, we can not use the Künneth formula. To avoid this difficulty we proceed as follows.

$k^\#$ is injective: Each element τ of $H^r(A^*(\pi W) \otimes A^*(S^p))$ can be written, by the "algebraic Künneth isomorphism" as $\tau = \{\alpha \otimes 1\} + \{\beta \otimes \eta\}$, where α is a closed r-V-form on πW, β is a closed (r-p)-V-form on πW and η is a representative of the orientation class of S^p. If $k^\#(\tau) = 0$, there is $\gamma \in A^r(W)$ such that $d\gamma = k(\alpha \otimes 1) + k(\beta \otimes \eta)$. We can regard the forms in the above identity as G-invariant forms on $S^p \times D$. By integrating this identity along the fibres in the trivial bundle $S^p \times D \longrightarrow D$ we see that β is an exact V-form on πW. From this it follows easily that α is also exact. This proves that $k^\#$ is injective.

$k^\#$ is exhaustive: The action of G on S^p beeing free we can think D as a subset of $S^p \times_G D$ of the form {point} x D.

Set $\widetilde{W} = (\{\text{point}\} \times D) \cap W$. For each $\{\alpha\} \in H^r(W)$ we can think α as a G-invariant closed form on $S^p \times \widetilde{W}$. Since $H^r(S^p \times \widetilde{W}) = (H^o(S^p) \otimes H^r(\widetilde{W})) \otimes (H^p(S^p) \otimes H^{r-p}(W))$ we can write $\{\alpha\} = \{1 \otimes \beta\} + \{\eta \otimes \gamma\}$ where β and γ are closed forms on $\widetilde{W}$ and η is a G-invariant orientation form on S^p. It is clear that β_G an γ_G can be regarded as closed forms on $\pi(W)$. A short computation shows that $k^{\#}(\{1 \otimes \beta_G\} + \{\eta \otimes \gamma_G\}) = \{\alpha\}$. Thus $k^{\#}$ is exhaustive q.e.d.

Let ∂ be the connecting morphism associated to the short exact sequence (1). Using Proposition 2 we define $\mathcal{D} = (\pi^{\#})^{-1} \circ \partial$ and obtain the following long exact sequence (the Gysin sequence of $\mathcal{F}$).

$$\longrightarrow H^r(B) \xrightarrow{\mathcal{D}} H^{r+p+1}(B) \xrightarrow{\pi^{\#}} H^{r+p+1}(M) \xrightarrow{\int_{\mathcal{F}}^{\#}} H^{r+1}(B)$$

As it is usual, we define now the Euler class of the sphere foliation $\mathcal{F}$ by $\chi_{\mathcal{F}} = \mathcal{D}(1) \in H^{p+1}(B)$, where $1 \in H^o(B)$

Remark. From the Gysin sequence we have that if the Euler class is zero (in particular, if p is even or $p + 1 > q$) then $H(M) = H(B) \otimes H(S^p)$. Also, if M is the sphere S^n $(n = p + q)$, using Poincaré's duality for V-manifolds (cf.[5]) and the Gysin sequence, we obtain $p = 2k - 1$ and $n = 2\,km - 1$.

3. Transverse Riemannian foliations

We recall that if $\mathcal{F}$ is a Hausdorff compact foliation on M there is a Riemannian metric g on M bundle-like with respect to $\mathcal{F}$. A codimension p foliation $\mathcal{F}^{\perp}$ on M transverse to $\mathcal{F}$ is said to be a Riemannian complementary of $\mathcal{F}$ if there is a Riemannian metric on M bundle-like both with respecto to $\mathcal{F}$ and $\mathcal{F}^{\perp}$.

Theorem 1. Let $\mathcal{F}$ be an oriented sphere foliation on a compact manifold M. If there exists a Riemannian complementary $\mathcal{F}^{\perp}$ to $\mathcal{F}$ the Euler class of $\mathcal{F}$ is zero. If $p = 1$ the converse is also true.

Proof. Let g be a Riemannian metric on M bundle-like with respect to the two foliations $\mathcal{F}$ and $\mathcal{F}^{\perp}$. We can cover M by local charts (U, x^i, y^{α}) such that the foliations $\mathcal{F}$ and $\mathcal{F}^{\perp}$ are given locally by x^i = constant and y^{α} = constant, respectively. In such a chart we have

$$g = g_{ij}(x)\, dx^i\, dx^j + g_{\alpha\beta}(y)\, dy^{\alpha}\, dy^{\beta}$$

Set

$$\eta = \sqrt{\det (g_{ij})} \quad dx^1 \wedge \ldots \wedge dx^p$$

where $x^1,\ldots,x^p$ are ordered posively respect to the orientation of $\mathcal{F}$. We have $d\eta = 0$, $h = \oint_{\mathcal{F}} \eta > 0$ and $dh = \oint_{\mathcal{F}} d\eta = 0$, so h is constant, different from zero, and the morphism $\oint^{\#}_{\mathcal{F}} : H^p(M) \longrightarrow H^o(B)$ is surjective. Thus the Euler class is zero.

If $p = 1$ we can find a Riemannian metric g and a unit Killing vector field ξ tangent to $\mathcal{F}$ (cf [7]). It can be shown that g is bundle-like with respect to $\mathcal{F}$. Without loss of generality we can assume g induces the same volume (= 1) on every regular leaf of $\mathcal{F}$ (see §4). Set $\theta(X) = g(\xi, X)$. As ξ is a Killing vector field we have $L_\xi \theta = 0$ and thus there exists $\tau \in A^2(B)$ such that $d\theta = \pi^*\tau$. Since $\oint_{\mathcal{F}} \theta = 1$, $\chi_{\mathcal{F}} = [\tau]$.

Therefore $\chi_{\mathcal{F}} = 0$ implies $\tau = d\gamma$. Set $\theta' = \theta - \pi^*\gamma$. Since $\theta'(\xi) = \theta(\xi) = 1$ and $d\theta' = 0$, θ' defines a foliation $\mathcal{F}^\perp$ transverse to $\mathcal{F}$ and of codimension 1. Let $\tilde{g}$ be the Riemannian metric on the V-manifold B induced by g. The Riemannian metric $g' = \theta' \otimes \theta' + \pi^*g$ is then bundle-like with respect to $\mathcal{F}$ and $\mathcal{F}^\perp$ q.e.d.

Remark. If $M = S^n$ the Euler class of the foliation is not zero and so $\mathcal{F}$ does not admit any Riemannian complementary.

§4. An integral formula

Recall (see Satake [6]) that on a compact orientable even dimensional Riemannian V-manifold B we have

$$\int_B \tilde{\Omega} = \chi_V(B)$$

where $\tilde{\Omega}$ is the Gauss-Bonnet integrand for V-manifolds and $\chi_V(B)$ is the Euler characteristic of B as V-manifold. (It should be noted that generally $\chi_V(B)$ does not coincide with the ordinary Euler characteristic and it is not necessarily an integer).

Theorem 2. Let M be a manifold of dimension $p + q$, endowed with a compact Hausdorf oriented foliation of codimension q. Then, there exists a bundle-like Riemannian metric g on M such that all regular leaves of $\mathcal{F}$ have the same volume, vol $\mathcal{F}$.

If M is compact and q even, we have

$$\frac{1}{vol\ F} \int_M f(\Omega_{ij}, F) = \chi_V\ (B)$$

where $f(\Omega_{ij}, F)$ is a function which depends on the curvature forms Ω_{ij} of g and on the foliation F.

Note that the right side of this formula is independent of the metric.

Proof. Let g' be a bundle-like metric on M and let ω' be the p-form on M associated to g', i.e., for each leaf L, the restriction of ω' to L is the volume element of the metric induced by g' on L. Let $h = f_F\ \omega'$. Then $g = (h^{-2/p} \circ \pi) g'$ is the required metric.

Then B inherits a Riemannian metric $\tilde{g}$, and from the Gauss-Bonnet theorem for V-manifolds we have

$$\int_M \pi^* \tilde{\Omega} \wedge \omega = \int_B (\tilde{\Omega} \wedge f_F\ \omega) = \mathrm{vol}\ F.\ \chi_V\ (B)$$

where ω is the p-form associated to g. Then, theorem follows from the fact that $\pi^*\tilde{\Omega}$ is directely computable from the curvature form Ω_{ij} of g.

Remark.- In the hypothesis of Duchamp's theorem [1, th.15.19], that is, there exists a complementary oriented foliation F' transversal to the above foliation F such that g is F'-bundle-like and flat respect to F, our integrand coincides with Duchamp's one. Thus, the theorem of Duchamp also holds for Hausdorff compact foliations, using $\chi_V(M/F)$ instead of the topological Euler characteristic of the leaf space.

REFERENCES

(1) Duchamp , T. Characteristic invariants of G. foliations. Illinois Thesis.

(2) Epstein, D.B.A. Foliations with all leaves compact. Ann. Inst. Four. (Grenoble), 26 (1976) 265-282.

(3) Greub, W., Halperin, S., Vanstone, R. Connections, Curvature and Cohomology. Academic Press, Inc.

(4) Haefliger, A. Some remarks on foliations with minimal leaves. J. Diff. Geometry, 15 (1980) 269-284.

(5) Satake, I. On a generalization of the notion of manifold. Proc. Nat. Acad. Sci. U.S.A. 42 (1957) 359-363.

(6) Satake, I. The Gauss-Bonnet theorem for V-manifolds. J. of Math. Soc. Japan. 9 (1957) 464-492.

(7) Wadsley, A.W. Geodesic foliations by circles. J. Diff. Geometry,10 (1975) 541-549.

NIJENHUIS TENSOR FIELD AND WEAKLY KAHLER MANIFOLDS

G. B. Rizza
Istituto di Matematica
Università. Via Università 12
43100 Parma. Italy

1.- INTRODUCTION

The essential role played by the Nijenhuis tensor field N in the problem of the integrability of an almost complex structure J is well known. The aim of the present paper is to point out further properties of the tensor field in the case when the manifold V is endowed with an almost hermitian structure.

An interesting relation linking the Nijenhuis field N and the field DJ, deduced from J by covariant differentation in the riemannian connection, is obtained in Sec. 4 (Th.1).

This relation is a very useful tool to prove a series of theorems.

A new proof of the known fact, that G_1-spaces and G_2-spaces can be defined in terms of N only, is given in Sec. 6 (Th.3).

A characterization of hermitian manifolds is also obtained in Sec. 6 (Th. 2). This result points out that there is a sort of analogy between these manifolds and G_1-spaces.

Finally, some known classes of almost hermitian manifolds can be characterized, by assigning particular expressions in terms of DJ to the tensor field N (Th. 4, Th. 5, Th. 6; Sec. 7).

2.- ISOMORPHISMS α, W, λ, γ

Let V be an almost hermitian manifold of dimension 2n and class C^{2n+1} ([1])

([1]) For the basic facts about almost hermitian manifolds see K. Yano [11], ch. 9; S. Kobayashi-K. Nomizu [5], II, ch. 9.

Let T^r_s the linear space of tensor fields of type (r,s) on V. In particular, let g be the symmetric field of T^0_2 of class C^1, defining the riemannian metric on V and let J be the field of T^1_1 of class C^{2n}, defining the almost complex structure on V.

Some isomorphisms of T^1_2 play an essential role in the following; namely α, W, λ, γ.

Let σ, ε be the homomorphisms of symmetry, of skew-symmetry of T^1_2; then $\alpha = \sigma - \varepsilon$. The isomorphisms W,λ are defined for any field L of T^1_2 by

$$WL = - c^1_3(c^2_2(L\otimes J)\otimes J) \quad , \quad \lambda L = c^1_3(L\otimes J) \qquad (^2)$$

Denote by G the symmetric tensor field of T^2_0 satisfying $c^1_1(g\otimes G) = \delta$ ([3]). Then the isomorphism γ is defined for any field L od T^1_2 by

$$\gamma L = c^2_2(c^1_1(g\otimes L)\otimes G)$$

Equivalent definitions of the isomorphisms α, λ, W are the following. Let L be an arbitrary field of T^1_2 ; then for any X, Y of T^1_0, we put

$$(\alpha L)(X,Y) = L(Y,X)$$

$$(\lambda L)(X,Y) = JL(X,Y) \quad , \quad (WL)(X,Y) = - JL(X,JY) \qquad (^4)$$

Similarly, the isomorphism γ can be implicitly defined by

$$g((\gamma L(X,Y),Z) = g(L(Z,Y),X)$$

where Z is an arbitrary field of T^1_0 and g(,) denotes inner product.

The above definitions shows that the isomorphisms W, λ, introduced in [6], depend only on the almost complex structure J and that the isomorphism γ, introduced in [7], depends only on the riemannian structure g.

The basic relations about the isomorphisms α, W, λ, γ are ([5]).

$$\alpha\alpha = \gamma\gamma = WW = 1 \quad , \quad \lambda\lambda = - 1 \qquad (1)$$

$$\alpha\lambda = \lambda\alpha \ , \ \alpha\gamma\alpha = \gamma\alpha\gamma \ , \ W\lambda = \lambda W \qquad (2)$$

([2]) The symbol c^r_s denotes contraction ([1], p. 45).

([3]) δ is the classical Kronecker field of T^1_1

([4]) Here J is regarded as an isomorphism of T^1_o.

([5]) See [6], n. 3,5 and [7], n. 3.

$$\alpha W\alpha W = W\alpha W\alpha = -\gamma W\gamma \qquad (3)$$

$$\lambda\gamma\lambda\gamma = \gamma\lambda\gamma\lambda = \alpha W\alpha \qquad (4)$$

We conclude the Section with two remarks

P_1 - <u>The isomorphisms γ and $\alpha W\alpha$ commute</u>.

P_2 - <u>The isomorphisms α, γ, W, λ are linked by relation</u>

$$(\alpha W\alpha)\lambda(\alpha\gamma\alpha) = (\alpha\gamma\alpha)\lambda(\alpha W\alpha) \qquad (5)$$

P_1 is an immediate consequence of (4), $(1)_1$. Using $(2)_2$, (4) and then (1), we can write

$$(\alpha W\alpha)\lambda(\alpha\gamma\alpha) = \gamma\lambda\gamma\lambda\lambda\gamma\alpha\gamma = -\gamma\lambda\alpha\gamma$$

$$(\alpha\gamma\alpha)\lambda(\alpha W\alpha) = \gamma\alpha\gamma\lambda\lambda\gamma\lambda\gamma = -\gamma\alpha\lambda\gamma$$

Since α and λ commute, we obtain P_2.

3.- TENSOR FIELD DJ

We denote by DJ the tensor field of T^1_2 obtained from J by covariant differentiation with respect to the Levi-Civita connection (riemannian connection) ([6]).

It is worth recalling some <u>relations</u> concerning DJ ([7])

$$WDJ = -DJ \qquad (6)$$

$$(W\gamma\lambda + \lambda\gamma)DJ = 0 \quad , \quad (W\lambda\gamma + \gamma\lambda)DJ = 0 \qquad (7)$$

$$(1 + \alpha\gamma\alpha)DJ = 0 \qquad (8)$$

We are able now to prove some propositions, we need in the following

P_3 - <u>The tensor field DJ satisfies relation</u>

$$(\alpha\gamma\alpha)\lambda DJ = \lambda(\alpha\gamma\alpha)DJ \qquad (9)$$

([6]) The index of covariant differentiation is assumed to be the first lower index.

([7]) For relations (6), (7) see [9], Sec. 3. Relation (8) is an immediate consequence of the skew-symmetry of the field $J = c^1_1(g\otimes J)$ of T^0_2 (Kähler form) (see [9], proposition P_1, p. 870).

P_4 - <u>The tensor field DJ satisfies relations</u>

$$(1 + \alpha\gamma\alpha)(1 - \alpha W\alpha)\lambda DJ = 0 \qquad (10)$$

$$(1 + \gamma)\alpha(1 - \alpha W\alpha)\lambda DJ = 0 \quad , \quad (1 + \alpha)\gamma(1 - \alpha W\alpha)\lambda DJ = 0 \qquad (11)$$

Using (6), $(2)_1$, $(1)_1$, we get

$$(\alpha\gamma\alpha)\lambda DJ = - \alpha\gamma\alpha\lambda WDJ = - \alpha\gamma\lambda(\alpha W\alpha)\alpha DJ$$

Since $\alpha W\alpha = \lambda\gamma\lambda\gamma$ ((4), Sec. 2), relation (9) follows immediately by virtue of (1), $(2)_1$.

To prove P_4, remark first that λ commutes with $\alpha W\alpha$ and with $1 - \alpha W\alpha$ ([2], Sec. 2). Then, applying relations (5), (9), we have

$$(1 + \alpha\gamma\alpha)(1 - \alpha W\alpha)\lambda DJ = (1 - \alpha W\alpha)\lambda(1 + \alpha\gamma\alpha)DJ$$

Now relation (10) follows immediately by virtue of (8). Finally, using the action of the isomorphisms α, γ on relation (10), we obtain relations (11).

We conclude this Section considering the <u>Kähler field</u>

$$K = (1 - \alpha - \gamma)DJ \qquad (^8) \qquad (12)$$

wich occurs in Sec. 5.

4.- NIJENHUIS TENSOR FIELD.

Consider the <u>Nijenhuis tensor field</u> N defined by

$$N = \frac{1}{2}\lambda\varepsilon W\varepsilon DJ \qquad (^9) \qquad (13)$$

It is worth recalling that

$$WN = - N \qquad (^{10}) \qquad (14)$$

Now, we are able to point out an interesting relation linking N with DJ . More explicitly

---

(8) If K denotes the differential of the Kähler form J, then $K = c_3^1(G\otimes K)$.

(9) Compare with [11], (1.10) p. 192 and note that our definition differs by a factor from the definition in [11].

(10) See [9], Sec. 3, $(9)_2$, p. 874.

Th. 1 - <u>For the tensor fields N, DJ, we have</u>

$$(1 - \alpha W \alpha) DJ = 4\lambda(1 + 2\sigma\gamma) N \tag{15}$$

Remark also that <u>relation</u> (15) <u>depends on the almost hermitian structure of</u> V.

To prove Th. 1, remember first that $2\varepsilon = 1 - \alpha$

Then, taking account of (6), $(1)_1$, from (13) we get

$$8N = \lambda(1 - \alpha) W (1 - \alpha) DJ = \lambda(W - \alpha W - W\alpha + \alpha W \alpha) DJ =$$

$$= - \lambda((1 - \alpha W\alpha) - \alpha(1 - \alpha W \alpha)) DJ = - \lambda(1 - \alpha)(1 - \alpha W \alpha) DJ$$

Since $2\sigma = 1 + \alpha$ and λ commutes with α, W ((2), Sec. 2), we can write

$$8(1 + 2\sigma\gamma) N = - (1 + \gamma + \alpha\gamma)(1 - \alpha)(1 - \alpha W \alpha)\lambda DJ =$$

$$= - ((1 - \alpha\gamma\alpha) - (\alpha + \gamma\alpha) + (\gamma + \alpha\gamma))(1 - \alpha W \alpha)\lambda DJ$$

Now, by virtue of relations (10), (11) (P_4, Sec. 3) we have

$$4(1 + 2\sigma\gamma) N = -(1 - \alpha W \alpha)\lambda DJ = - \lambda(1 - \alpha W \alpha) DJ$$

and, since $\lambda\lambda = -1$ ($(1)_2$, Sec. 2), we obtain relation (15).

5.- WEAKLY KAHLER MANIFOLDS.

Many classes of almost Hermitian manifolds, generalizing the class of Kähler manifolds, are known in the literature ([11])([12]). Some of them occur in the following Sections.

Consider first G_1-spaces and G_2-spaces, introduced by L. Hervella and E. Vidal in [4]. G_1-spaces, also known as <u>underkähler manifolds</u>, can be defined by

$$\sigma(1 - \alpha W \alpha) DJ = 0 \qquad ([13]) \tag{16}$$

([11]) See the classification of A. Gray and L. M. Hervella in [3] and also S. Sawaki [10], G. B. Rizza [8], S. Donnini [2], where other classes have been introduced and studied.

([12]) For the manifolds belonging to such classes we may use the general name "weakly Kähler manifolds".

([13]) Compare with [3], p. 41, taking account of (6) of Sec. 3.

G_2-spaces can be defined by

$$(1 - \alpha W\alpha)K = (\alpha W - W\alpha)DJ \qquad (17)$$

where K is the Kähler field introduced in Sec. 3.

To prove relation (17), remark that a known condition defining G_2-spaces ([3], p. 41) can be written in the form

$$DJ - \alpha DJ - \gamma DJ + \alpha W\alpha WDJ + W\alpha DJ + \alpha W\alpha\gamma DJ = 0 \qquad (18)$$

Thus, taking account of (12), (6), $(1)_1$, we obtain (17).

Consider then almost Kōto manifolds, called also quasi-Kähler manifolds or almost O^*-spaces ([14]). The defining condition is

$$(1 + \alpha W\alpha)DJ = 0 \qquad ([15]) \qquad (19)$$

Finally it is worth recalling that almost Tachibana manifolds (nearly Kähler manifolds) and almost Kähler manifolds can be defined by

$$\sigma DJ = 0 \quad , \quad K = 0 \qquad (20)$$

respectively.

6.- First results.

We want to show now that new and old results, concerning weakly Kähler manifolds, can be derived very easily, by using relation (15) of theorem 1.

Th. 2 - A necessary and sufficient condition in order that V be a Hermitian manifold is

$$\varepsilon(1 - \alpha W\alpha)DJ = 0 \qquad (21)$$

Th. 3 - If V is a G_1-space, a G_2-space, then, respectively

$$\sigma\gamma N = 0 \quad , \quad N = 2\varepsilon\gamma N \qquad (22)$$

and conversely.

([14]) See [3], p. 40; [11], p. 197-198.
([15]) Compare with [11], (4.1) p. 197.

Theorem 2 improves the known result stating that condition $(1 - \alpha W\alpha)DJ =$ is equivalent to $N = 0$ ([16]) and also points out, that there exits a sort of analogy between G_1-spaces and Hermitian spaces ([17]).

Theorem 3 is known ([18]). It shows that G_1-spaces and G_2-spaces can be defined in terms of the tensor field N, torsion of the almost complex structure.

To prove the above theorems, remark first that from relation (15) of Sec. 4 we get

$$\sigma(1 - \alpha W\alpha)DJ = 8\lambda\sigma\gamma N \quad , \quad \varepsilon(1 - \alpha W\alpha)DJ = 4\lambda N \quad ([19]) \qquad (23)$$

Since λ is an isomorphism, from (23) we derive immediately Th. 2 and the part of Th. 3 concerning G_1-spaces.

To complete the proof of Th. 3, we must show that conditions (17), $(22)_2$ are equivalent. By virtue of (6), relation (18), equivalent to (17) of Sec. 5, can be written in the form

$$- 4\varepsilon W\varepsilon DJ = (1 - \alpha W\alpha)\gamma DJ \qquad (24)$$

Taking into account of (13), $(1)_2$ and of the property P_1 of Sec. 2, we see that (24) is equivalent to

$$8\lambda N = \gamma(1 - \alpha W\alpha)DJ \qquad (25)$$

Now, as $\alpha W\alpha N = - N$ and $\lambda\alpha W\alpha = - \gamma\lambda\gamma$ ([20]), using relation (15) of Sec. 4, we can write equation (25) in the form

$$2\gamma\lambda\gamma N = \gamma\lambda(1 + 2\sigma\gamma)N \qquad (26)$$

Finally, since $\gamma\lambda$ is an isomorphism (Sec. 2), (26) results to be equivalent to $(22)_2$ ([21]).

([16]) Compare with K. Yano [11], (2.6), p. 193.

([17]) Compare (16), (21).

([18]) Compare with A. Gray-L. M. Hervella [3], p. 41.

([19]) By virtue of $(2)_1$ λ commutes with σ and with ε. N is a skew-symmetric tensor field.

([20]) The first equation follows immediately from (14) and from the skew-symmetry of N, the second one is an obvious consequence of (4), $(1)_2$.

([21]) Note that $1 = \sigma + \varepsilon$.

7.- CHARACTERIZATION THEOREMS.

We are able now to show that some classes of weakly Kähler manifolds can be characterized by particular expressions for the tensor field N. More explicitly, we have the theorems

Th. 4 - *If the Nijenhuis field has the form*

$$N = -\frac{1}{4}\lambda(1-\alpha W\alpha)DJ \quad , \quad N = -\frac{1}{8}(\gamma\lambda+\lambda\gamma)DJ \tag{27}$$

then V is respectively a G_1-space, a G_2-space; and conversely.

Th. 5 - *If the Nijenhuis field has the form*

$$N = -\frac{1}{2}\lambda\varepsilon DJ \tag{28}$$

then V is an almost Köto manifold (almost O^*-space); and conversely.

Th. 6 - *If the Nijenhuis field has the form*

$$N = -\frac{1}{2}\lambda DJ \quad , \quad N = -\frac{1}{4}\lambda\gamma DJ \tag{29}$$

then V is respectively an almost Tachibana manifold, an almost Kähler manifold; and conversely.

The converse proposition of the result about almost Tachibana manifolds is known ([22]).

From $(29)_1$, $(29)_2$ some known relations concerning the invariants can be immediately deduced ([23]).

It is also worth remarking that the fundamental relation of Th. 1 in Sec. 4 plays an essential role in the proofs of the theorems.

([22]) Compare [11], (4.16), p. 142 and take into account the remark of the footnote ([9]).

([23]) See [3], p. 54. Remark only that for any tensor field T of T^1_2 we have $\|\lambda T\| = \|\gamma T\| = \|T\|$.

8.- PROOFS

By virtue of Th. 3, to prove the first statement of Th. 4 it suffices to show that $(22)_1$ and $(27)_1$ are equivalent. This fact however is an immediate consequence of relation (15) of Sec. 4 ([24]).

Since $\lambda\gamma$ is an isomorphism, taking account of (4), (1) of Sec. 2, we can write relation (15) in the form

$$(\lambda\gamma + \gamma\lambda)DJ = 4\ \lambda\gamma\lambda(1 + 2\sigma\gamma)N$$

It is now immediate to derive that $(22)_2$ and $(27)_2$ are equivalent ([25]). Therefore, by virtue of Th. 3, the second statement of Th. 4 is also proved.

We prove now Th. 5. Since λ is an isomorphism, by virtue of (14) from (28) we derive $W\varepsilon DJ = -\ \varepsilon DJ$. Hence we have

$$\varepsilon DJ - W\varepsilon DJ + 2\varepsilon W\varepsilon DJ = 0$$

and this relation is equivalent to (19) ([26]). Conversely, we know that relation (15) implies relations (23) of Sec. 6. Using $(23)_2$, from (19) we immediately get (28).

To prove the first statement of Th. 6, remark that, since $\sigma N = 0$ and λ commutes with σ, from $(29)_1$ we get $(20)_1$. Conversely, since an almost Tachibana manifold is an almost Köto manifold ([27]), from $(20)_1$, (28) we derive $(29)_1$.

Finally, we prove the second statement of Th. 6. By virtue of (14), from $(29)_2$ we obtain $W\gamma DJ = -\ \gamma DJ$. Then, by (6), (3), $(1)_1$ we can write

$$\alpha W\alpha DJ = -\ \alpha W\alpha WDJ = \gamma W\gamma DJ = -\ \gamma\gamma DJ = -\ DJ$$

So V is an almost Köto manifold. Comparing $(29)_2$ with (28), we immediately derive $K = 0$ ([28]). Conversely, since an almost Kähler manifold is an almost Köto manifold ([29]), from (12), $(20)_2$, (28) we obtain $(29)_2$.

We can now conclude that all the theorems of Sec. 7 have been proved.

([24]) Only remember $(1)_2$.
([25]) Just use relation (4) and note that $\alpha W\alpha N = -\ N$.
([26]) Use (6) of Sec. 3 and equation $\alpha = 1 - 2\varepsilon$.
([27]) See [11], Th. 4.5, p. 199.
([28]) See (12) in Sec. 3.
([29]) See [11], Th. 4.4, p. 199.

REFERENCES

[1] Bourbaki N. , Algebra 3, Hermann, Paris, 1968.

[2] Donnini S. , Due generalizzazioni delle varietà quasi Kähleriane, Riv. Mat. Univ. Parma, 4, 1978, p. 485-492.

[3] Gray A. - Hervella L. M. , The sixteen classes of almost hermitian manifolds and their linear invariants, Ann. di Mat. 123, 1980, p. 35-58.

[4] Hervella L. M. - Vidal E. , Nouvelles géométries pseudo-kählerien-nes G_1 et G_2, C. R. Acad. Sc. Paris, 283, 1976, p. 115-118.

[5] Kobayashi S. - Nomizu K. , Foundations of differential geometry , (I, II), Interscience Publ., New York, 1963, 1969.

[6] Rizza G. B. , Teoremi di rappresentazione per alcune classi di connessioni su di una varietà quasi complessa, Rend. Ist. Mat. Univ. Trieste, 1, 1969, p. 9-25.

[7] Rizza G. B. , Connessioni metriche sulle varietà quasi hermitiane, Rend. Ist. Mat. Univ. Trieste, 1, 1969, p. 163-181.

[8] Rizza G. B. , Varietà parakähleriane, Ann. di Mat., 98, 1974, p. 47-61.

[9] Rizza G. B. , Almost complex conditions and weakly Kähler manifolds, Riv. Mat. Univ. Parma, 5, 1979, p. 869-877.

[10] Sawaki S. , On almost-hermitian manifolds satisfying a certain condition on the almost-complex structure tensor, Diff. Geom. in honor of K. Yano, Kinokuniya, Tokio, 1972, p. 443-450.

[11] Yano K. , Differential geometry on complex and almost complex spaces Pergamon Press, Oxford, 1965.

GENERIC EMBEDDINGS, GAUSS MAPS AND STRATIFICATIONS

M.C. Romero-Fuster
University of Southampton
Universidad de Valencia

0.- Introduction

Let M be a closed oriented smooth $(= C^\infty)$ m- manifold and f a smooth embedding of M into $\mathbb{R}^{m+s}$. Any vector $v \in S^{m+s-1}$ defines a smooth height function f_v on M by putting $f_v(x) = \langle f(x), v\rangle$, for all $x \in M$. As M is compact the absolute minimum for the function f_v on M is attained at one or more points of M. In the generic cases we can define two different stratifications of S^{m+s-1} by considering : 1) the set of all the critical points of each height function on M (Gauss stratification), or 2) the set of absolute minima of f_v, for all $v \in S^{m+s-1}$ (core stratification). These stratifications provide us of 'good picture' of the distribution of the generic singularities of the Gauss map $\gamma(f)$ on M. We give some results in this direction for the case $s = 1$, $m \leqslant 6$, and generalize some of them for $s > 1$. Moreover, the core stratification comes out to be tightly related, by means of the Gauss map of M induced by f, to the structure of the boundary $H(f)$ of the convex hull $CH(f)$ of fM in $\mathbb{R}^{m+s}$, (we give here some results, see [9] for a wider treatment of this subject). We finish our study with some applications to curves, surfaces and 3-manifolds.

Acknowledgements : I would like to thank S.A. Robertson for helpful comments and suggestions valuable to the production of this work.

Contents :

5. Consequences in lower dimensions.

1. The Gauss stratification

Let us consider the inner product function

$$\lambda(f) : M \times S^{m+s-1} \longrightarrow \mathbb{R}$$
$$(x,v) \longmapsto \langle f(x),v\rangle = f_v(x)$$

This is a smooth function and it induces a continuous map

$$\Lambda(f) : S^{m+s-1} \longrightarrow C^\infty(M)$$
$$v \longmapsto f_v$$

where $C^\infty(M)$ denotes the space of smooth functions on M.

In [7, pp. 39, 40, 41] Looijenga constructed a natural Whitney stratification $\mathcal{S}^k(M)$ of the k-jet space $J^k(M)$ of M off the subset $W^k(M)$ composed by the k-jets of which some representative is not k-determined. (For $m \leqslant 6$ and $k \leqslant 8$ this stratification coincides with that defined by the simple orbits of codimension $\leqslant 6 + m$ under the action of the group $G^8(M,\mathbb{R}) = G^8(M) \times G^8(\mathbb{R})$, where $G^8(X)$ denotes the group of 8-jets of diffeomorphisms of X). By using the stratifications $\mathcal{S}^k(M)$, $k \in \mathbb{Z}^+$, Looijenga also constructed, [7, §8], a stratification $\mathcal{S}(M)$ of $C^\infty(M)\setminus W(M)$, $W(M)$ being a subset of infinite codimension composed of those functions $f \in C^\infty(M)$ having a non-algebraically isolated critical point. Let us give a sketch of Looijenga's construction.

Given a parameters manifold P and a smooth family of functions $F : M \times P \longrightarrow \mathbb{R}$ on M parametrized by P, let us denote by $j^k_* F : M \times P \longmapsto J^k(M)$ the k-jet extension of F, (i.e. $j^k_*F(x,p) = j^k F_p(x)$).

<u>1.1 Definition</u>.- Let $(S,\mathcal{S})$ be a finite stratification of a subspace S of $J^k(M)$. A family of functions $F : M \times P \longrightarrow \mathbb{R}$ is said to be <u>multitransverse with respect to $(S,\mathcal{S})$</u> if

a) $j^k_*F : M \times P \longrightarrow J^k(M)$ is transverse to $\mathcal{S}$.

b) $D(F)|_{(j_*^kF)^{-1}(S,\mathcal{S})} : (j_*^kF)^{-1}(S,\mathcal{S}) \longrightarrow \mathbb{R} \times P$ has regular intersections relative to $\pi : \mathbb{R} \times P \longrightarrow P$, where

$$D(F) : M \times P \longrightarrow \mathbb{R} \times P$$
$$(x,p) \longmapsto (F(x,p),p) .$$

<u>1.2 Lemma</u> ([7] p. 49) . Given $\eta \in C^\infty(M) \setminus W(M)$ let V be a 'slice' throughout the $\mathrm{Diff}(M) \times \mathrm{Diff}_c(\mathbb{R})$ -orbit of η in $C^\infty(M)$, where $\mathrm{Diff}_c(\mathbb{R})$ is the group of diffeomorphisms of $\mathbb{R}$ with a compact support. Let's define a family of functions on M parametrized by V , $F ; M \times V \longrightarrow \mathbb{R}$ by $F(x,\eta') = \eta'(x)$. Then there is a neighbourhood V_o of η in V such that $F_o = F|M \times V_o$ defines a transversal family (i.e. there is some k such that F_o is multitransverse with respect to $\mathcal{S}^k(M)$ and $j_*^kF_o : M \times V_o \longrightarrow J^k(M)$ avoids $W^k(M)$). (For definition and existence of 'slices' see [13]).

Let $(\mathcal{S}_o,\mathcal{S}_o',\mathcal{S}_o'')$ be the canonical stratification (in Looijenga's sense) of the deformation

$$\begin{array}{ccccc} M \times V_o & \xrightarrow{D(F_o)} & \mathbb{R} \times V_o & \xrightarrow{\pi_{V_o}} & V_o \\ (x,\xi) & \longmapsto & (\xi(x),\xi) & \longmapsto & \xi \end{array}$$

associated to F_o ; $(\mathcal{S}_o,\mathcal{S}_o',\mathcal{S}_o'')$ stratify the spaces $M \times V_o$, $\mathbb{R} \times V_o$ and V_o respectively).

Observe that two functions η_1 and η_2 of V_o are in the same stratum if and only if

a) η_1 and η_2 have the same number of critical points of a given type; and

b) η_1 and η_2 have the same number of critical values with given multiplicities.

The stratification $\mathcal{S}_o''$ of V_o induces a stratification $\mathcal{S}_\eta$ on a neighbourhood U_η of η in $C^\infty(M)$, U_η being formed by fattening up the slice V_o with suitable neighbourhoods of the points of V_o in the corresponding $\mathrm{Diff}(M) \times \mathrm{Diff}_c(\mathbb{R})$ - orbits. Since the various pairs

$(\mathcal{S}_\eta, U_\eta)$ agree on overlaps, they induce a stratification $\mathcal{S}(M)$ of $C^\infty(M) \setminus W(M)$ whose strata are invariant under the action of $\mathrm{Diff}(M) \times \mathrm{Diff}_c(\mathbb{R})$.

<u>1.3 Definition</u>.- Given an embedding $f \in \mathrm{Emb}(M, \mathbb{R}^{m+s})$ we say that f is <u>Λ-generic</u> if $\Lambda(f) : S^{m+s-1} \longrightarrow C^\infty(M)$ is transverse to the strata of $\mathcal{S}(M)$ and avoids $W(M)$.

Given $f \in \mathrm{Emb}(M, \mathbb{R}^{m+s})$, let $S(\mathbb{R}^{m+s+1})$ be the unit sphere in $\mathbb{R}^{m+s} \times \mathbb{R}$, and define a function

$$G_f : M \times S(\mathbb{R}^{m+s+1}) \longrightarrow \mathbb{R}$$
$$(x,(p,t)) \longmapsto t\|f(x)\|^2 - 2\langle p, f(m)\rangle$$

<u>1.4 Definition</u>.- An embedding f is said to be <u>distance-generic</u> if both G_f and its restriction to the sphere $S_o(\mathbb{R}^{m+s+1})$ defined by $t = 0$ are transverse to $\mathcal{S}(M)$ and avoid $W(M)$.

<u>1.5 Theorem</u>.- (Looijenga [7]). The subset of distance-generic embeddings is open and dense in $\mathrm{Emb}(M, \mathbb{R}^{m+s})$.

From this we get immediately the following,

<u>1.6 Corollary</u>.- The subset of Λ-generic embeddings is dense in $\mathrm{Emb}(M, \mathbb{R}^{m+s})$.

We shall denote the subset of Λ-generic embeddings of $\mathrm{Emb}(M, \mathbb{R}^{m+s})$ by $\Lambda(M, \mathbb{R}^{m+s})$.

Given any $f \in \Lambda(M, \mathbb{R}^{m+s})$, the stratification $\mathcal{S}(M)$ pulls back by $\Lambda(f)$ to a Whitney regular stratification $\tilde{\mathcal{G}}(f)$ of S^{m+s} .

<u>1.7 Definition</u>.- The <u>Gauss stratification</u> of S^{m+s-1} is the stratification $\mathcal{G}(f)$ composed of the connected components of the strata of $\tilde{\mathcal{G}}(f)$.

Notice that $\mathcal{G}(f)$ satisfies the frontier condition (i.e. the boundary of each stratum is a union of strata of lower dimension), for it is made of the connected components of a Whitney regular stratification, (see [4]) .

2. The core stratification

2.1 Definition.- The Maxwell subset $Mxw(M)$ of the function space $C^\infty(M)$ is defined as the closed subset of $C^\infty(M)\setminus W(M)$, such that any function ξ on its complement $(C^\infty(M)\setminus W(M))\setminus Mxw(M)$ attains its absolute minimum at a unique non-degenerate critical point, (see [14]).

2.2 Definition.- Given $f \in Emb(M,\mathbb{R}^{m+s})$, we define the Maxwell subset of S^{m+s-1} associated to f to be the subset

$$Mxw(f) = \{v \in S^{m+s-1} \mid f_v \in Mxw(M) \subset C^\infty(M)\}$$

2.3 Notation.- Let $f \in C^\infty(M)$ be a function reaching its absolute minimum at r different points $p_1,\dots,p_r$ of M. Suppose that the germ of f at p_i has a singularity of type A_{k_i} (for suitable k_i in Arnold's notation [1]). Then we say that f has an absolute minimum of type $A_{k_1} + \dots + A_{k_r}$. Note that for a Λ-generic embedding f, the points $v \in Mxw(f)$ may correspond to height functions $f_v \in C^\infty(M)$ with absolute minimum type $A_{2j_1+1} + \dots + A_{2j_r+1}$ with $r - 1 + \sum_{i=1}^{r} 2j_i \leqslant m+s-1$, (since for a function ξ to have an absolute minimum at a point x, the germ of ξ at x must have a singularity of type A_{2j+1} at x for some $j \geqslant 0$ when $m \leqslant 6$.

We shall suppose throughout this section that the embedding f is Λ-generic and $m \leqslant 6$.

Following Looijenga we know that $Mxw(M)$ is a union of strata of $\mathcal{S}(M)$, and hence $Mxw(f)$ is a union of strata of $\mathcal{G}(f)$. We want to construct a new stratification of $Mxw(f)$ by joining together strata of $\mathcal{G}(f)|Mxw(f)$ in such a way that any of the new strata will be the largest connected submanifold of S^{m+s-1} all whose points correspond to height functions with the same type of absolute minimum.

Let us write

$$\mathcal{A}_{j_1,\dots,j_r} = \{v \in S^{m+s-1} \mid f_v \text{ has its absolute minimum of type } A_{2j_1+1} + \dots + A_{2j_r+1}\}$$

2.4 Lemma.- Each of the subsets $\mathcal{A}_{j_1,\dots,j_r} \subset Mxw(f)$ is a submanifold of S^{m+s-1} with codimension $r - 1 + \sum_{i=1}^{r} 2j_i$. Moreover, each of these submanifolds is a union of strata of $\mathcal{G}(f)$.

A proof is given in [11].

2.5 Theorem.- The partition $\mathcal{A}(f)$ of Mxw(f) into the connected components of the subsets $\{\mathcal{A}_{j_1,\dots,j_r} \mid r - 1 + \sum_{i=1}^{r} 2j_i \leq m + s - 1\}$ is a Whitney regular stratification of Mxw(f).

Proof: Regularity of the elements of $\mathcal{A}(f)$ follows easily from regularity of $\mathcal{G}(f)$.

2.6 Definition.- The core stratification of S^{m+s-1} associated to the embedding f is the one whose strata are the elements of $\mathcal{A}(f)$ together with the connected components of $S^{m+s-1} \setminus Mxw(f)$. We denote this stratification by $\mathcal{C}(f)$.

The stratification $\mathcal{C}(f)$ is Whitney regular and satisfies the frontier condition.

2.7 Remark.- The strata of $\mathcal{C}(f)$ can be classified into the following types.

1) Morse strata = connected components of $S^{m+s-1} \setminus Mxw(f)$;

2) conflict strata = connected components of $\mathcal{A}_{0,\overset{r \text{ times}}{\dots},0}$, $1 < r \leq m + s$;

3) bifurcation strata = connected components of $\mathcal{A}_j$, $j > 0$;

4) mixed strata = connected components of $\mathcal{A}_{j_1,\dots,j_r}$, $r > 1$ and at least one $j_k > 0$, $1 \leq k \leq r$.

We shall denote the submanifolds defined by the union of conflict, bifurcation or mixed strata of codimension k in S^{m+s-1}, respectively by $C_k(f)$, $B_k(f)$ or $M_k(f)$. Observe that $C_k(f) = \mathcal{A}_{0,\overset{(k+1) \text{ times}}{\dots},0}$; $B_{2k}(f) = \mathcal{A}_k$ and $M_k(f) = \{\mathcal{A}_{j_1,\dots,j_r} \mid r > 1,$ some $j_k > 0$ and $r - 1 + \sum_{i=1}^{r} 2j_i = k\}$.

We now study the relation among the Euler numbers of all these strata.

2.8 Theorem.- Let $\mathcal{S} = \{S_j\}_{j=o}^{n}$, $n < \infty$ be a Whitney regular stratification of a closed subset S of a manifold M , satisfying the frontier condition. Then the Euler number e(M) of the manifold M is given by

$$e(M) = e(M \setminus S) + \sum_{j=o}^{n} (-1)^{\operatorname{codim} S_j} e(S_j) .$$

A proof of this result can be found in [11].

Let us apply this to the closed subset Mxw(f) of S^{m+s-1} with the stratification $\mathcal{A}(f)$, then we get immediately,

2.9 Corollary.- For any $f \in \Lambda(M,\mathbb{R}^{m+s})$ the following equality holds

$$e(S^m \setminus Mxw(f)) + \sum_{j=o}^{m} (-1)^j \{e(B_j(f)) + e(M_j(f)) + e(C_j(f)) =$$

$$= \begin{cases} 2 \text{ if } m \text{ even} \\ 0 \text{ if } m \text{ odd} \end{cases} .$$

3. Gauss maps and convex hulls

Let us consider $f \in Emb(M,\mathbb{R}^{m+1})$, then the Gauss map $\gamma(f) : M \longrightarrow S^m$ associated to f is defined by sending each point $x \in M$ to the inward unit normal to fM at f(x) translated to the origin of $\mathbb{R}^{m+1}$.

Let CH(M) be the convex hull of fM in $\mathbb{R}^{m+1}$, i.e. CH(M) is the intersection of all the convex subsets of $\mathbb{R}^{m+1}$ that contain M . Let us denote by H(M) the boundary of CH(M) . Observe that H(M) is a C^1 hypersurface of $\mathbb{R}^{m+1}$. Let $\Gamma(f)$ denote the Gauss map on H(M) .

3.1 Definition .- The set of exposed points E(f) of f is the compact subset $fM \cap H(f)$.

Let $\tilde{C}_j = \gamma(f)^{-1}(C_j(f)) \cap E(f)$. Notice that $\tilde{C}_j$ is the set of points of fM at which there is a supporting hyperplane with j + 1 points of contact, (remember that $C_j(f)$ is the union of all the conflict strata of $\mathcal{C}(f)$ with codimension j in S^m).

Let M_v represent the (finite) subset of absolute minima of f_v. Note that $M_v = \gamma(f)^{-1}(v) \cap E(f)$ and also that $M_v \cap M_w \neq \emptyset \Longrightarrow v = w$.

<u>3.2 Theorem</u>.- There exists an open and dense subset $I(m)$ of $\Lambda(M, \mathbb{R}^{m+1})$ such that M_v is in general position in $\mathbb{R}^{m+1}$, for all $v \in S^m$. (Proof : [9]).

<u>3.3 Corollary</u>.- Let P_j be the convex hull of $\tilde{C}_j$ in $\mathbb{R}^{m+1}$. Then $\Gamma(f)|P_j : P_j \longrightarrow C_j(f)$ is a fiber bundle with fiber a closed j-simplex with exactly $j+1$ vertices.

<u>Proof</u>: Observe that given any $v \in C_j(f)$, from above theorem it follows that the convex hull H_v of M_v is a j-simplex with exactly $j+1$ vertices. Let $\Delta_j = \langle p_o, \cdots, p_j \rangle$ denote the standard j-simplex in $\mathbb{R}^{j+1}$. Then given any $v \in C_j(f)$, there is a homeomorphism

$$\xi_v : H_v \longrightarrow \Delta_j$$

$$x = \sum_{i=o}^{j} \xi_i(x) v_i \longmapsto \xi_v(x) = \sum_{i=o}^{j} \xi_i(x) p_i$$

where $\{\xi_i(x)\}_{i=o}^{j}$ are the barycentric coordinates of x in $H_v = \langle v_o, \cdots, v_j \rangle$. Now, the ξ_v vary continuously with v. Then the homeomorphism

$$P_j \longrightarrow C_j(f) \times \Delta_j$$

$$x \longmapsto (\Gamma(f)(x), \xi_{\Gamma(f)(x)}(x))$$

provides a global chart for the map $\Gamma(f)|P_j : P_j \longrightarrow C_j(f)$ showing that it is, in fact, a (trivial) topological fiber bundle projection.

There are similar results for the submanifolds $M_j(f)$ and $B_j(f)$, so that it can be seen that $H(M)$ may be covered by disjoint fiber bundles of the above type, (see [9]).

<u>4. Local situation: singularities of Gauss maps and dual Gauss maps</u>

<u>4.1 Definition</u>.- The <u>dual Gauss map</u> $\tau(f) : M \longrightarrow G(m)$ associated to an embedding $f \in \mathrm{Emb}(M, \mathbb{R}^{m+1})$ is defined by sending each point

$x \in M$ to the point defined by the oriented tangent space to fM at $f(x)$ in the affine Grassmannian $G(m)$ of oriented hyperplanes in $\mathbb{R}^{m+1}$.

Observe that $x \in M$ is a critical point of $\tau(f)$ if and only if x is a critical point of $\gamma(f)$, if and only if x is a degenerate critical point of $f_{\pm\gamma(f)(x)}$.

By using the transversality conditions on Λ-generic embeddings and standard results of the theory of singularities of smooth functions, [15], [17], one gets the following

<u>4.2 Proposition</u> .-Let $f \in \Lambda(M, \mathbb{R}^{m+1})$, $m \leqslant 6$, then for all $x \in M$ and for all $v \in S^m$, the germ of the inner product function $\lambda(f) : M \times S^m \longrightarrow \mathbb{R}$ at (x,v) is a versal unfolding of the germ of f_v at x.

(Proof : [3], [10], [11]).

<u>4.3 Corollary</u>.- Given $f \in \Lambda(M, \mathbb{R}^{m+1})$, $m \leqslant 5$ we have

a) the only singularities of $\gamma(f) \in C^\infty(M, S^m)$ are elementary catastrophes,

b) the germ of the image-set of $\tau(f)$ at any of the critical points of $\tau(f)$ is equivalent to the level bifurcation set of an elementary catastrophe.

<u>Proof</u> : Let $\nu(f) : M \longrightarrow G(1)$ be defined by attaching to each point $x \in M$ the point given by the oriented normal line to fM at $f(x)$ in the affine Grassmannian $G(1)$ of oriented lines in $\mathbb{R}^{m+1}$. The map $\nu(f)$ is a local diffeomorphism. Observe that, locally, we can identify $G(1)$ with the product $U \times V$, U being an open subset of M (or $\mathbb{R}^m$) and V an open subset of S^m (or $\mathbb{R}^m$). We can also (locally) identify $G(m)$ with $V \times \mathbb{R}$. Then we get the following commutative diagram,

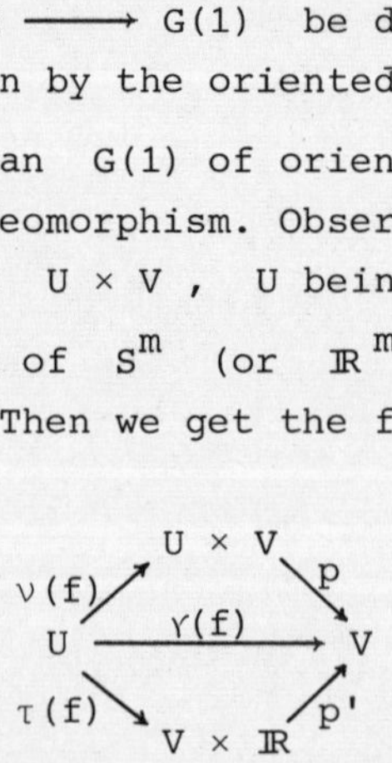

p and p' being the obvious projections.

On the other hand we have that $\lambda(f) : U \times V \longrightarrow \mathbb{R}$ defines a versal unfolding at each point. Moreover, the image set $\mathrm{Im}\nu(f)$ of the map $\nu(f)$ can be identified with the catastrophe manifold of $\lambda(f)$. Hence p is a catastrophe map and so must be $\gamma(f)$ (being equivalent to p by means of $\nu(f)$), proving a).

For part b), observe that we can choose local coordinates for which $\tau(f)(x) = (\gamma(f)(x), \lambda(f)(x,\gamma(f)(x)))$, then use the definition of level bifurcation set of an elementary catastrophe, (see [1]).

<u>4.4 Remark</u>.- A different study of part a) in which stable Gauss maps are identified with stable lagrangian projections can be found in [16]. That the images of stable dual Gauss maps are stable Legendre submanifolds was observed by V.M. Zakalyukin in [18].

<u>4.5 Corollary</u>.- Given $f \in \Lambda(M,\mathbb{R}^{m+1})$, $m \leqslant 6$ then

a) the singularities of $\gamma(f)$ lying on $E(f)$ are Boardman singularities of type $S_{1,\ldots,1,0}$ or $S_{1,\ldots,1}$;

b) If $\gamma(f)$ has a singularity of type $S_{\underbrace{1,\ldots\ldots\ldots}_{k\text{ times}},1,0}$ at x then $\tau(f)$ has a singularity of type $S_{\underbrace{1,\ldots\ldots\ldots\ldots\ldots}_{(k+1)\text{ times}},1}$ at x.

<u>Proof</u> : See [5] for definition of Boardman singularities and observe that the only singularities of $\gamma(f)$ on $E(f)$ must be absolute minima of height functions.

We now make a brief analysis of the Differential Geometry of the embedded manifold in neighbourhoods of some of the various critical points. We can write

$$d\gamma(f)(x) = \begin{pmatrix} \gamma_1(f) & & 0 \\ & \ddots & \\ 0 & & \gamma_m(f) \end{pmatrix}(x) ,$$

where $\gamma_j(f) \in C^\infty(M)$, $j = 1,\ldots,m$, are the principal curvatures of fM at f(x). Let $e_j(x) \in C^\infty(M,TfM)$, $j = 1,\ldots,m$ be the principal directions of curvature. We have that $x \in S_1(\gamma(f))$ (notation as in [5]) if and only if $\mathrm{corank}\, d\gamma(f)(x) = 1$. We can then suppose without loss of generality that in a neighbourhood U of x

$$S_1(\gamma(f)) \cap U = \{x \in U \mid \gamma_1(f)(x) = 0 \text{ and } \gamma_j(f)(x) \neq 0 \ \forall j \neq 1\}$$

moreover by shrinking U if necessary we may assume that the $\gamma_j(f)$, $j \neq 1$ do not vanish on U. In all the generic cases we consider (at least for $m \leq 5$) $S_1(\gamma(f))$ defines a submanifold of U of dimension $\leq m - 1$. Any $x \in S_1(\gamma(f))$ will be either in $S_{1,1}(\gamma(f))$ or $S_{1,0}(\gamma(f))$ according to $e_1(x)$ is in $T_{f(x)}S_1(\gamma(f))$ or not. This defines again a submanifold $S_{1,1}(\gamma(f))$ of $S_1(\gamma(f))$ of dimension $\leq m - 2$ and succesively, further submanifolds $S_{1,1,\dots,1}(\gamma(f))$ may be defined by looking at the positions of the vectors $e_1(x)$. Notice also that $S_1(\gamma(f))$ separates in U a region in which $\gamma_1(f) > 0$ from another one in which $\gamma_1(f) < 0$. The sign of the Gauss curvature, $K(f) = \det(d\gamma(f))$, in either of these regions will depend on the signs of the remaining $\gamma_j(f)$ which do not change on U, but in any case it will be opposite in one region to the other.

In general we can say that at a critical point x of $\gamma(f)$ of type S_p, exactly p of the principal curvatures vanish ($\gamma_1(f)(x) = \cdots = \gamma_p(f)(x) = 0$, $\gamma_j(f)(x) \neq 0$, $j > p$). Now, provided $S_p(\gamma(f))$ is a submanifold, we have that the corresponding principal directions of curvature ($e_j(x)$, $1 \leq j \leq p$) are transversal to this submanifold at any point of type $S_{p,0}$, and q of them are tangent to it at the $S_{p,q}$ - points, $0 < q \leq p$.

Let us now consider the case of higher codimensional submanifolds, that is $s > 1$. First of all we need to give an analogue of the Gauss map for this case. This is done as follows. Given $f \in \mathrm{Emb}(M, \mathbb{R}^{m+s})$ let

$$\sum fM = \{(x,v) \in M \times S^{m+s-1} \mid v \perp T_{f(x)} fM\}$$

be the unit normal bundle of M in $\mathbb{R}^{m+s}$. The set $\sum fM$ has a standard manifold structure, (see [6]), and it may be identified with the boundary of a tubular neighbourhood of fM in $\mathbb{R}^{m+s}$. We can consider the two following projections with domain in $\sum fM$.

1) The projection $\xi : \sum fM \longrightarrow M$, $(p,v) \longmapsto p$, which defines an

$(s-1)$-sphere bundle whose fibre $\sum_p fM$ over any point $p \in M$ is the $(s-1)$-sphere of unit vectors in $\mathbb{R}^{m+s}$ which are orthogonal to $T_{f(p)}fM$.

2) The projection $G(f) : \sum fM \longrightarrow S^{m+s-1}$

$$(p,v) \longmapsto v \quad .$$

<u>4.6 Definition</u>.- The above projection $G(f)$ is called the Gauss map induced by f.

<u>4.7 Proposition</u>.- The singular set of the function $f_v \in C^\infty(M)$ is the subset $\xi(G(f)^{-1}(v)) \subset M$.

<u>Proof</u> : See [8].

<u>4.8 Proposition</u>.- The following are equivalent,

i) the Gauss map $G(f)$ has a singularity at a point $(p,v) \in \sum fM$,

ii) p is a degenerate (or non-Morse) singular point of f_v,

iii) the second fundamental form II_v of f in the direction v is degenerate at p.

<u>Proof</u> : See [8].

Transversality conditions on $\Lambda(f)$ for Λ-generic embeddings apply also in this case to give :

<u>4.9 Proposition</u>.- Let $f \in \Lambda(M, \mathbb{R}^{m+s})$, $m + s \leqslant 5$, then the germ of $\lambda(f)$ at (p,v) is a versal unfolding of the germ of f_v at p, for all $p \in M$, for all $v \in S^{m+s-1}$.

And by an analogous argument to the above one for $s = 1$ we get

<u>4.10 Corollary</u>.- Given any $f \in \Lambda(M, \mathbb{R}^{m+s})$, $m + s \leqslant 5$, the singularities of the Gauss map $G(f) : \sum fM \longrightarrow S^{m+s-1}$ induced by f, are equivalent to elementary catastrophes.

Moreover, we can define (locally) the dual Gauss map induced by f by

$$T(f) : \sum fM \longrightarrow S^{m+s-1} \times \mathbb{R}$$
$$(p,v) \longmapsto (v, f_v(p)) \quad .$$

Then we immediately get,

4.11 Corollary.- The map $T(f)$ has a critical point at $(p,v) \in \sum fM$ if and only if (p,v) is a critical point of $G(f)$. Furthermore, the germ of the image set of $T(f)$ at a critical point (p,v) is equivalent to the level bifurcation set germ corresponding to the elementary catastrophe defined by $G(f)$ at (p,v).

5. Consequences in lower dimensions

a) $m = 1$, $s = 1$: closed oriented curve embedded in $\mathbb{R}^2$.

In the generic case we may have inflexion points of the curve, corresponding to fold points $(S_{1,0})$ of $\gamma(f)$ and cusps $(S_{1,1})$ of $\tau(f)$.

b) $m = 2$, $s = 1$: closed oriented surface in $\mathbb{R}^3$.

Generically, the Gauss map $\gamma(f)$ may have fold and cusp points ($S_{1,0,0}$ and $S_{1,1,0}$) and correspondingly $\tau(f)$ will have swallowtail $(S_{1,1,1})$ and cusp points. For example, consider a twisted torus given by the parametric equations

$$\left.\begin{array}{l} x = (A + B\cos\phi)\cos\theta \\ y = (A + B\cos\phi)\sin\theta \\ z = B\sin\phi + \sin 2\theta \end{array}\right\} \qquad A > B > 0 \text{, constants.}$$

The zero-contour $K_o = K(f)^{-1}(0)$ of the Gauss curvature $K(f)$ defines two curves of critical points of $\gamma(f)$ and $\tau(f)$. In these curves there are eight points $\{x_i\}_{i=1}^{8}$ such that in any sufficiently small neighbourhood U_i of each x_i we have the following situation: the curve $K_0 \cap U_i$ separates U_i in two regions, one of negative curvature and another of positive curvature. One of these regions contains a curve of self-intersections points of $\tau(f)$, which is tangent to $K_0 \cap U_i$ at x_i. The points x_i are the cusp points of $\gamma(f)$, whereas the rest of the points in K_o are fold points (see [2] for details and more examples).

We also have the following result

5.1 Corollary.- Let $f \in \Lambda(M, \mathbb{R}^3)$ then the following equality holds

$$\alpha - \beta = 4 - 2e(\overset{\circ}{\widehat{E(f)}})$$

where

α = number of cusp points of $\gamma(f)$ lying on $E(f)$

β = number of supporting planes of M with triple points of contact.

$\overset{\circ}{\widehat{E(f)}}$ = interior of $E(f)$ in fM.

<u>Proof</u> : It follows from corollary 2.9 by observing that

$B_2(f) = \gamma(f)(\{\text{cusp points of } \gamma(f) \text{ lying on } E(f)\})$,

$B_j(f) = \emptyset$ for all $j \neq 2$,

$M_j(f) = \emptyset$ for all j ,

$C_2(f) = \{\text{critical values of multiplicity three of } \gamma(f)|E(f)\}$

$C_1(f) = \{\text{critical values of multiplicity two of } \gamma(f)|E(f)\}$

$C_j(f) = \emptyset$ for all $j \neq 1, 2$.

Notice that $B_2(f)$ and $C_2(f)$ are finite sets of points, and $C_1(f)$ is a finite disjoint union of curves which may be either closed or ending at points in $B_2(f) \cup C_2(f)$. Then compute the different Euler numbers to get the required result (observe that $e(\overset{\circ}{\widehat{E(f)}}) =$ $= e(S^m \setminus Mxw(f)))$.

c) $m = 1$, $s = 2$: closed curve embedded in $\mathbb{R}^3$.

Generically we can avoid selfintersections ,(see [12]).

Let $\mu(s)$ and $\delta(s)$ represent the curvature and the torsion of the given curve $f : M = S^1 \longrightarrow \mathbb{R}^3$ at a point $f(s)$. We also have generically : 1) $\mu(s) \neq 0$ for all s, and 2) if $\delta(s) = 0$ then $\delta'(s) \neq 0$, (see [12]) .

We can embed the unit normal bundle of M in $\mathbb{R}^3$ by means of the parametrization

$$F : S^1 \times S^1 \longrightarrow \mathbb{R}^3$$
$$(s,q) \longmapsto f(s) + \varepsilon(\cos q P_s + \sin q B_s) ,$$

where P_s and B_s are the principal normal and binormal of the curve at $f(s)$ and ε is a small enough positive number (to avoid self-intersections of F). Then the image of F is diffeomorphic to $\sum fM$.

The Gauss map is given by

$$G(f) : \mathrm{Im}\,F \longrightarrow S^2$$
$$F(s,q) \longmapsto \cos q\, P_s + \sin q\, B_s \quad .$$

The generic singularities of $G(f)$ are folds $(S_{1,0,0})$ and cusps $(S_{1,1,0})$. Moreover we have

a) $F(s,q)$ is a critical point of $G(f)$ $\Longleftrightarrow$ $G(f) \circ F(s,q) = \pm B_s$,

b) $F(s,q)$ is a cusp point of $G(f)$ $\Longleftrightarrow$ $\delta(s) = 0$.

(For a proof of these two statements see [2]).

Hence we can say that a point $(p,v) \in \sum fM \subset S^1 \times S^2$ is a critical point of $G(f)$ iff $v = B_p$.

We also have the dual Gauss map

$$T(f) : \mathrm{Im}\,F \longrightarrow G(1) \equiv S^2 \times \mathbb{R} \quad \text{(locally)}$$
$$F(s,q) \longmapsto (G(f)(F(s,q)), \langle F(s,q), G(f)(F(s,q))\rangle) \quad .$$

The critical points of $T(f)$ are of the form $f(s) + \varepsilon B_s$ and their image by $T(f)$ is a parallel plane to the osculating plane of f at $f(s)$.

d) $m = 3$, $s = 1$: closed 3- manifolds in $\mathbb{R}^4$.

In the generic case the Gauss map is allowed to have swallowtail points $(S_{1,1,1,0})$ as well as elliptic and hyperbolic umbilic points $(S_{2,0})$. Corresponding to a swallowtail point of $\gamma(f)$, the map $\tau(f)$ will have a butterfly point $(S_{1,1,1,1})$ and in a neighbourhood of $S_{2,0}$ points of $\gamma(f)$, the image of $\tau(f)$ will look like their corresponding level bifurcation sets. On $E(f) \subset fM$ no points of type $S_{2,0}$ for $\gamma(f)$ will occur.

We now give some results for surfaces and 3-manifolds concerning the Gaussian curvature.

<u>5.2 Corollary</u>.- Let M be a closed oriented surface, then there exists an open and dense subset $\mathcal{K}_2$ of $\mathrm{Emb}(M,\mathbb{R}^3)$ such that for all $f \in \mathcal{K}_2$, the zero-contour $K(f)^{-1}(0)$ of the Gaussian curvature function $K(f) : M \longrightarrow \mathbb{R}$ is a submanifold of M.

<u>Proof</u> : Put $\mathcal{K}_2 = \Lambda(M,\mathbb{R}^3)$, we have that the singular set $\Sigma(\gamma(f))$ of $\gamma(f)$ is precisely $S_1(\gamma(f)) = S_{1,0}(\gamma(f)) \cup S_{1,1}(\gamma(f))$ which is a submanifold of M in the cases that can occur generically. But $K(f)^{-1}(0) = \Sigma(\gamma(f))$.

<u>5.3 Corollary</u>.- Let M be a closed oriented 3-manifold, then there exists an open and dense subset $\mathcal{K}_3$ of $\mathrm{Emb}(M,\mathbb{R}^4)$ such that for all $f \in \mathcal{K}_3$, $K(f)^{-1}(0)$ is a submanifold off a finite set of points of M. The latter is composed of the critical points of $\gamma(f)$ of type $S_{2,0}$. Moreover,

i) x is a hyperbolic umbilic point of $\gamma(f) \Longrightarrow x$ is a nondegenerate (Morse) critical point of $K(f)$ of index 2;

ii) x is an elliptic umbilic point of $\gamma(f) \Longrightarrow x$ is a nondegenerate critical point of $K(f)$ of index 1.

<u>Proof</u> : Put $\mathcal{K}_3 = \Lambda(M,\mathbb{R}^3)$. Then $K(f)^{-1}(0) = \Sigma(\gamma(f)) = S_1(\gamma(f)) \cup S_2(\gamma(f))$ (disjoint union). Now, by looking at the local models of elementary catastrophes, [15], one can immediately see that $\Sigma(\gamma(f))$ is a submanifold at all its points except at $S_2(\gamma(f))$. Moreover, $S_2(\gamma(f))$ must be composed of isolated umbilic points. The finiteness of $S_2(\gamma(f))$ follows from the compactness of $S_2(\gamma(f))$. Assertions i) and ii) follow from the commutativity of the diagram of corollary 4.3,

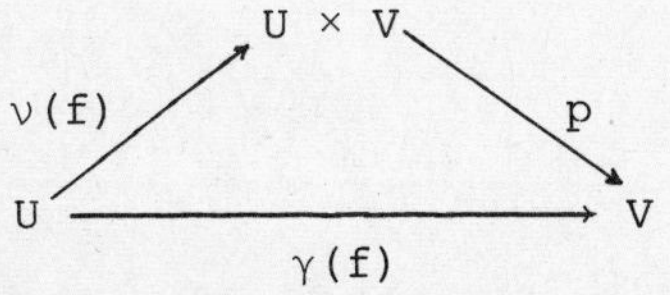

together with the fact that $p|\mathrm{Im}\,\nu(f)$ is a representative of an elementary catastrophe, and a straightforward calculation in suitable local coordinates.

REFERENCES

[1] Arnold, V.I. : "Critical points of smooth functions and their normal forms", Russian Math. Surveys 30 : 5 (1975), 1-75.

[2] Banchoff, T. , Gaffney, T. and McCrory, C : "Cusps of Gauss Mappings", Research Notes in Mathematics, Pitman 1982.

[3] Bruce, J.W. : "Duals of generic hypersurfaces". Math.Scand. (1981), 36 - 60.

[4] Gibson, C.G., Wirthmüller, K. , du Plessis, A.A. and Looijenga, E.J.N. : "Topological stability of smooth mappings", Lect. Notes in Math. 552, Springer, Berlin (1976).

[5] Golubitsky, M. and Guillemin, V. : "Stable Mappings and their Singularities". Grad. Texts in Math. 14, Springer, Berlin (1973).

[6] Hirsch, M.W. : "Differential Topology", Grad. Texts in Math. 33, Springer, Berlin (1976).

[7] Looijenga, E.J.N. : "Structural Stability of smooth families of C^∞-functions", Doctoral thesis, University of Amsterdam (1974).

[8] Nimmo-Smith, M.I. : "Curvatures, singularities of projections and smooth maps", Ph. D. thesis, University of Durham (1971).

[9] Robertson, S.A. and Romero Fuster, M.C. : "Convex hulls of hypersurfaces", (to appear).

[10] Romero Fuster, M.C. : "The convex hull of an immersion", Ph. D. thesis, Southampton University (1981).

[11] Romero Fuster, M.C. : "Sphere stratifications and the Gauss map", (to appear).

[12] Sedykh,V.D. : "Singularities of the convex hull of a curve in $\mathbb{R}^3$", Funct. Anal. Appl. 11 (1977).

[13] Sergeraert, F. "Un théorème des fonctions implicites pour les espaces de Fréchet", Ann. Sc. de l'Ec. Norm. Sup. t5 Fasc. 4 (1972), 559-660.

[14] Thom, R. : "Sur le cut-locus d'une variété plongée", J. Diff. Geom. 6 (1972), 577-586.

[15] Trotman D.J.A. and Zeeman E.C. : "The classification of elementary catastrophes of codimension $\leqslant$ 5", Structural Stability, the Theory of Catastrophes and Applications in the Sciences, Seattle 1975. Lect. Notes in Math. 525, Springer, Berlin (1976).

[16] Wall,C.T.C. :"Geometric properties of generic differentiable manifolds", Geometry and Topology, Rio de Janeiro 1976, Lect. Notes in Math. 597, Springer, Berlin (1977), 707-774.

[17] Wasserman,G. : "Stability of unfoldings", Lect. Notes in Math.

393, Springer, Berlin (1974).

[18] Zaralyukin,V.M. : "Singularities of convex hulls of smooth manifolds", Funct. Anal. Appl. 11 (1978).

SPECTRAL GEOMETRY OF SUBMANIFOLDS IN THE COMPLEX PROJECTIVE SPACE

A. Ros
Departamento de Geometría y Topología
Universidad de Granada.
Spain.

1.-INTRODUCTION.

The complex projective space can be isometrically embedded in a Euclidean space with parallel second fundamental form. S. S. Tai, [3], give a simple version of this embedding.

Let $\mathbb{C}P^m$ be the complex projective space with Fubini-Study metric of constant holomorphic sectional curvature 1. Any n-dimensional submanifold M^n of $\mathbb{C}P^m$ can be viewed as a submanifold of the Euclidean space. Let Δ be the Laplace-Beltrami operator of M acting on functions and λ_1 the first eigenvalue of this operator. Using certain results in spectral geometry of submanifolds in the Euclidean space we have obtained the following [2]:

Theorem 1.- Let M^{2n} be a compact complex submanifold of $\mathbb{C}P^m$. Then $\lambda_1 \leqslant n+1$. The equality holds if and only if M is totally geodesic.

Theorem 2.- Let M^p be a compact totally real minimal submanifold of $\mathbb{C}P^m$. a) If there exists a totally geodesic complex submanifold, $\bar{M}^{2p}$, of $\mathbb{C}P^m$ such that M^p is a totally real submanifold of $\bar{M}^{2p}$, then $(p+1)/2$ is an eigenvalue of Δ on M^p.
b) If $\lambda_1 = (p+1)/2$, then there exists a totally geodesic complex submanifold $\bar{M}^{2p}$ of $\mathbb{C}P^m$, such that M^p is a totally real submanifold of $\bar{M}^{2p}$.

2.-THE TAI'S EMBEDDING.

Let $HM(m) = \{A \in gl(m,\mathbb{C}) \ / \ \bar{A} = A^t\}$ the space of m×m-Hermitian matrices. We define on HM(m+1) the metric

$$g(A,B) = 2 \text{ trace } AB, \qquad \text{for all } A,B \text{ in } HM(m+1). \tag{1}$$

Let $\mathbb{C}P^m = \{A \in HM(m+1) \ / \ AA = A, \text{ trace } A = 1\}$. Then $\mathbb{C}P^m$ is a submanifold of HM(m+1), isometric to the complex projective space of constant holomorphic sectional curvature 1. Tangent and normal space at a point A of $\mathbb{C}P^m$ are given by

$$T_A(\mathbb{C}P^m) = \{X \in HM(m+1) \; / \; XA+AX = X\}, \qquad (2)$$

$$T_A^{\perp}(\mathbb{C}P^m) = \{Z \in HM(m+1) \; / \; AZ = ZA\}. \qquad (3)$$

If J is the complex structure of $\mathbb{C}P^m$, σ the second funfamental form of $\mathbb{C}P^m$ in HM(m+1) and $\tilde{H}$ the mean curvature vector, then

$$JX = \sqrt{-1}\,(I - 2A)X, \qquad (4)$$

$$\sigma(X,Y) = (XY + YX)(I - 2A), \qquad (5)$$

$$\sigma(JX,JY) = \sigma(X,Y), \qquad (6)$$

$$\tilde{H} = (1/2m)[I - (m+1)A], \qquad (7)$$

for all X,Y in $T_A(\mathbb{C}P^m)$, being I the $(m+1)\times(m+1)$-identity matrix. Finally we consider the relations

$$g(\sigma(E_1,E_1), \sigma(E_1,E_1)) = 1, \qquad (8)$$

$$g(\sigma(E_1,E_1), \sigma(E_2,E_2)) = \frac{1}{2}, \qquad (9)$$

where $E_1, E_2 \in T_A(\mathbb{C}P^m)$, $g(E_1,E_1) = g(E_2,E_2) = 1$ and $g(E_1,E_2)=g(E_1,JE_2)=0$.

3.-SUBMANIFOLDS IN THE COMPLEX PROJECTIVE SPACE.

Let M^n be a n-dimensional submanifold of $\mathbb{C}P^m$. Let $\{E_i\}_{i=1...n}$ an orthonormal base in the tangent space of M^n at certain point A, and H the mean curvature vector of M^n in HM(m+1). If M^n is minimal in $\mathbb{C}P^m$ then

$$H = (1/n)\sum_i \sigma(E_i,E_i). \qquad (10)$$

Hence, from (8), (9) and (10) we obtain the following

Lemma. Let M^{2n} (resp. M^p) a complex (resp. totally real minimal) submanifold of $\mathbb{C}P^m$ and H the mean curvature vector of M in HM(m+1). Then

$$g(H,H) = (n+1)/2n \qquad (11)$$

$$(\text{resp. } g(H,H) = (p+1)/2p\,). \qquad (12)$$

PROPOSITION. Let M^{2n} (resp. M^p) a complex (resp. totally real minimal) submanifold of $\mathbb{C}P^m$. Then M is minimal in some sphere of HM(m+1) if and only if M^{2n} is totally geodesic (resp. if and only if there exists a totally geodesic complex submanifold $\bar{M}^{2p}$ of $\mathbb{C}P^m$ such that M^p is a totally real submanifold of $\bar{M}^{2p}$). Moreover the radius of the sphere is $\sqrt{2n/(n+1)}$ (resp. $\sqrt{2p/(p+1)}$) .

Outline of the proof. Let M^{2n} a complex submanifold of $\mathbb{C}P^m$. If M is minimal in some sphere, we can show that exists a linear subspace $\mathbb{C}P^q$ of $\mathbb{C}P^m$, such that M^{2n} is a complex submanifold of $\mathbb{C}P^q$, and M^{2n} is minimal in some sphere of HM(q+1) whose center is of the type aI, where a is a real number and I is the $(q+1)\times(q+1)$-identity matrix. Let H the mean curvature vector of M in HM(q+1). Then $H = h\cdot(A - aI)$, being h a non zero real number. From (11) and the well know relation $g(H,A - aI) = -1$, we obtain

$$g(A - aI,\ A - aI) = 2n/(n+1). \tag{13}$$

On the other hand

$$g(A - aI,\ A - aI) = 2(q+1)a^2-4a+2\ . \tag{14}$$

From (13) and (14) we obtain n = q. Hence M^{2n} is open in $\mathbb{C}P^q$, so that M^{2n} is totally geodesic in $\mathbb{C}P^m$.

Using (7) we conclude the converse.

By a similar method, take into account of (6), we obtain the result for totally real submanifolds.

Proof of theorems 1 and 2. For a compact submanifold M^n of the Euclidean space, B.Y. Chen ,[1] ,has proved the following inequality

$$\int_M \alpha^n \geqslant \left(\frac{\lambda_1}{n}\right)^{n/2} \mathrm{vol}(M),$$

where $\alpha = \sqrt{g(H,H)}$ is the mean curvature of M and vol (M) denotes the volume of M. If the equality holds, then M is minimal in some sphere of the Euclidean space. In this case from a well know result of T. Takahashi n/(radius of the sphere) is an eigenvalue of Δ on M. Now we conclude easily the proof.

REFERENCES

[1] B.Y.CHEN, Geometry of submanifolds and its applications. Science University of Tokyo, 1981

[2] A. ROS, Spectral geometry of CR-minimal submanifolds in the complex projective space. Kodai Math. J. (to appear).

[3] S.S.TAI, Minimum embedding of compact symmetric spaces of rank one, J. Diff. Geometry, 2, 1968, 55-66.

SELF-DUAL AND ANTI-SELF-DUAL HOMOGENEOUS STRUCTURES

F. Tricerri and L. Vanhecke

In this paper we start with a brief survey on the theory of homogeneous Riemannian structures. Then we concentrate on some special features of four-dimensional Riemannian manifolds. We introduce the notion of self-dual and anti-self-dual homogeneous structures and finally we give some examples. We refer to [6] for more details and further information.

1. GENERAL HOMOGENEOUS RIEMANNIAN STRUCTURES

As is well-known E. Cartan proved that a connected, complete and simply connected Riemannian manifold is a symmetric space if and only if the curvature is constant under parallel translation. Ambrose and Singer extended this theory in order to be able to characterize homogeneous Riemannian manifolds by a local condition which is to be satisfied at all points.

THEOREM 1.1 [1] *Let* (M,g) *be a connected, complete and simply connected Riemannian manifold with Levi Civita connection* ∇ *and Riemann curvature tensor* R. *Then* (M,g) *is homogeneous, i.e. there exists a transitive and effective group* G *of isometries of* M, *if and only if there exists a tensor field* T *of type* $(1,2)$ *such that*

$$(1.1)\qquad \begin{cases} \text{i)}\ g(T_XY,Z) + g(Y,T_XZ) = 0 , \\ \text{ii)}\ (\nabla_X R)_{YZ} = [T_X, R_{YZ}] - R_{T_XYZ} - R_{YT_XZ} , \\ \text{iii)}\ (\nabla_X T)_Y = [T_X, T_Y] - T_{T_XY} , \end{cases}$$

for $X,Y,Z \in \mathfrak{X}(M)$, *or equivalently, with* $\tilde{\nabla} = \nabla - T$:

$$(1.2)\qquad \begin{cases} \text{i')}\ \tilde{\nabla} \textit{ is a metric connection} , \\ \text{ii')}\ \tilde{\nabla}R = 0 , \\ \text{iii')}\ \tilde{\nabla}T = 0 . \end{cases}$$

In their paper Ambrose and Singer set up a natural correspondence between the solutions of the system (1.1) and the groups G acting transitively and effectively on M as a group of isometries. Moreover they suggest the possibility of classifying Riemannian homogeneous manifolds by properties of the tensor field T.

Note that there may exist more than one solution of the system (1.1) on the same manifold. Therefore we consider the following definitions.

DEFINITION 1.2. A *homogeneous (Riemannian) structure* on (M,g) is a tensor field T of type (1,2) which is a solution of the system (1.1).

DEFINITION 1.3. Let T and T' be homogeneous structures on (M,g). Then T and T' are said to be *isomorphic* if and only if there exists an isometry φ of (M,g) such that

$$\varphi_*(T_X Y) = T'_{\varphi_* X}\varphi_* Y , \qquad X,Y \in \mathfrak{X}(M) .$$

This notion of isomorphism is very natural in the classification problem because isomorphic homogeneous structures give rise to the same group of isometries.

Next we give a brief sketch of a kind of algebraic classification. Let p be a point of M and let $V = T_pM$. V is a Euclidean vector space over $\mathbb{R}$ with inner product $< , >$ induced from the metric g on M. In what follows we will consider tensors T of type (0,3) instead of tensors of type (1,2).

Let $\mathcal{T}(V)$ be the vector subspace of $\overset{3}{\otimes}V^*$, V^* being the dual of V, determined by all the (0,3)-tensors having the same symmetries as a homogeneous structure, i.e.

$$\mathcal{T}(V) = \{T \in \overset{3}{\otimes}V^* | T_{xyz} + T_{xzy} = 0 , \qquad x,y,z \in V\} .$$

$\mathcal{T}(V)$ is a Euclidean vector space with inner product defined by

$$< T,T' > = \sum_{i,j,k} T_{e_ie_je_k} T'_{e_ie_je_k} ,$$

where $\{e_1,\dots,e_n\}$ is an arbitrary orthonormal basis of V. Further, there is a natural action of the orthogonal group O(V) on $\mathcal{T}(V)$ given by

$$(aT)_{xyz} = T_{a^{-1}x a^{-1}y a^{-1}z}$$

for $x,y,z \in V$ and $a \in O(V)$.

Next put

$$c_{12}(T)(z) = \sum_i T_{e_ie_iz} , \qquad z \in V ,$$

and define the subspaces $\mathcal{T}_i(V)$, $i = 1,2,3$, of $\mathcal{T}(V)$ by

$$\mathcal{T}_1(V) = \{T \in \mathcal{T}(V) \mid T_{xyz} = \langle x,y \rangle \varphi(z) - \langle x,z \rangle \varphi(y) \ , \ \varphi \in V^*\} \ ,$$

$$\mathcal{T}_2(V) = \{T \in \mathcal{T}(V) \mid \mathfrak{S}_{xyz} T_{xyz} = 0 \ , \quad c_{12}(T) = 0\} \ ,$$

$$\mathcal{T}_3(V) = \{T \in \mathcal{T}(V) \mid T_{xyz} + T_{yxz} = 0\} \ ,$$

for $x,y,z \in V$. $\mathfrak{S}$ denotes the cyclic sum. Then we have

THEOREM 1.4. *For* $\dim V \geqslant 3$, $\mathcal{T}(V)$ *is the orthogonal direct sum of the subspaces* $\mathcal{T}_i(V)$, $i = 1,2,3$. *Moreover, these spaces are invariant and irreducible under the action of* $O(V)$. *Further, for* $\dim V = 2$, *we have* $\mathcal{T}(V) = \mathcal{T}_1(V)$ *where* $\mathcal{T}(V)$ *is irreducible. Finally, when* $n = \dim V$, *we have*

$$\dim \mathcal{T}(V) = \frac{n^2(n-1)}{2} \ , \qquad \dim \mathcal{T}_1(V) = n \ ,$$

$$\dim \mathcal{T}_2(V) = \frac{n(n-2)(n+2)}{3} \ , \quad \dim \mathcal{T}_3(V) = \binom{n}{3} \ .$$

Hence there are, in general, eight invariant subspaces, the trivial spaces included. These considerations lead to

DEFINITION 1.5. Let $\mathcal{J}(V)$ be an invariant subspace of $\mathcal{T}(V)$. A homogeneous structure T on (M,g) is said to be of *type* $\mathcal{J}$ when $T(p) \in \mathcal{J}(T_pV)$ for all $p \in M$.

So we may consider eight classes of homogeneous structures. Note that the type of a homogeneous structure is invariant under isomorphisms of homogeneous structures.

We refer to [6] for examples of the eight different types of homogeneous structures and for further properties and results. There we also give other characterizations for some of these classes. For example, we prove that (M,g) is a *naturally reductive homogeneous manifold* if and only if there exists a homogeneous structure T on (M,g) of type $\mathcal{T}_3$. This theorem is used extensively to study the geometry of the *generalized Heisenberg groups* and the remarkable geometry of the six-dimensional example in relation with the theory of harmonic, commutative and D'Atri spaces. The study of these spaces provided a motivation for our interest in the theorem of Ambrose and Singer.

2. FOUR-DIMENSIONAL GEOMETRY

The study of the four-dimensional case takes a special and important place in Riemannian geometry. This is very well illustrated by the theory of *self-duality* and *anti-self-duality* (see for example [2]). This special feature is due to the fact that the rotation group SO(4) is not simple but locally isomorphic to SU(2) × SU(2).

It is for this reason that the space of curvature tensors $\mathcal{R}(V)$ over a four-dimensional real vector space V with inner product has an extra decomposition. In general $\mathcal{R}(V)$ splits into three irreducible invariant subspaces under the action of the orthogonal group O(4) :

$$\mathcal{R}(V) = \mathcal{R}_1(V) \oplus \mathcal{W} \oplus \mathcal{R}_2(V).$$

But in dimension four the conformal invariant part $\mathcal{W}$ decomposes further into

$$\mathcal{W} = \mathcal{W}_+ \oplus \mathcal{W}_-$$

under the action of SO(4) (see [2],[5]). Further, let W denote the projection of a curvature tensor $R \in \mathcal{R}(V)$ on $\mathcal{W}$, i.e. W is the *Weyl tensor*. Then one defines : An oriented four-dimensional Riemannian manifold is *self-dual* (or *anti-self-dual* respectively) if its Weyl tensor $W = W_+$ (or $W = W_-$ respectively), i.e. if $W_- = 0$ (or $W_+ = 0$ respectively).

In what follows we consider the space $\mathcal{T}(V)$ when dim V = 4 and study the decomposition under the action of the special orthogonal group. In the first place we note that

$$\dim \mathcal{T}_1(V) = 4 \ , \ \dim \mathcal{T}_2(V) = 16 \ , \ \dim \mathcal{T}_3(V) = 4.$$

Next, let Λ^2 denote the space of exterior 2-forms on V. Then we have

$$\mathcal{T}(V) = V^* \otimes \Lambda^2 .$$

V is equipped with an inner product $<,>$ and in what follows we fix an orientation on $(V,<,>)$. Then the Hodge star operator $* : \Lambda^2 \longrightarrow \Lambda^2$ is defined by

$$*\lambda \wedge \mu = (\lambda,\mu)\omega \in \Lambda^4$$

where $\lambda,\mu \in \Lambda^2$. (λ,μ) denotes the induced inner product of the two-forms λ,μ and ω is the volume form defined by $<,>$ and the orientation. Note that $*$ is a symmetric linear operator such that $*^2 = 1$. Then Λ^2 splits into a direct sum

$$\Lambda^2 = \Lambda^2_+ \oplus \Lambda^2_-$$

where $\Lambda^2_\pm$ are the ± 1 eigenspaces of $*$. A two-form of Λ^2_+ is called *self-dual* and a two-form of Λ^2_- is called *anti-self-dual*.

Now we return to the space $\mathcal{T}(V)$. Since $\mathcal{T}_1(V) \cong V^*$ and $\mathcal{T}_3(V) \cong V^*$, these two spaces are also irreducible under the action of SO(4) but it follows from the general theory (see for example [7]) or from what we remarked above, that $\mathcal{T}_2(V)$ splits further into two irreducible components. It is not difficult to prove the following.

THEOREM 2.1. *Let* V *be an oriented four-dimensional real vector space with inner product* $<,>$. *Then we have the orthogonal direct sum*

$$\mathcal{T}(V) = \mathcal{T}_1(V) \oplus \mathcal{T}_2^+(V) \oplus \mathcal{T}_2^-(V) \oplus \mathcal{T}_3(V)$$

where the summands are irreducible invariant subspaces under the action of SO(4). *Moreover,*

$$\dim \mathcal{T}_2^+(V) = \dim \mathcal{T}_2^-(V) = 8$$

and

$$\mathcal{T}_2^+(V) = \{T \in \mathcal{T}_2(V) \mid T_{x*(yz)} = T_{xyz}\} \quad ,$$

$$\mathcal{T}_2^-(V) = \{T \in \mathcal{T}_2(V) \mid T_{x*(yz)} = - T_{xyz}\} \ , \quad x,y,z \in V.$$

The projections of $T \in \mathcal{T}(V)$ *on these four spaces are given by*

$$p_1(T)_{xyz} = < x,y > \varphi(z) - < x,z > \varphi(y)$$

where

$$\varphi(z) = \frac{1}{3} c_{12}(T)(z) ;$$

$$p_3(T)_{xyz} = \frac{1}{3} \mathop{\mathfrak{S}}_{x,y,z} T_{xyz} ;$$

$$p_2^+(T)_{xyz} = \frac{1}{2} \{p_2(T)_{xyz} + p_2(T)_{x*(yz)}\} ,$$

$$p_2^-(T)_{xyz} = \frac{1}{2}\{p_2(T)_{xyz} - p_2(T)_{x*(yz)}\}$$

where

$$p_2(T) = T - p_1(T) - p_3(T)$$

and $x,y,z \in V$.

This theorem leads to

DEFINITION 2.2. Let T be a homogeneous structure on an oriented four-dimensional Riemannian manifold. Then T is said to be *self-dual* (or *anti-self-dual* respectively) if $p_2^-(T) = 0$ (or $p_2^+(T) = 0$ respectively).

In the next section we shall give an example of a homogeneous Riemannian manifold with a self-dual and an anti-self-dual homogeneous structure. Further research is needed to construct examples for the sixteen classes in order to be able to decide about the inclusion relations between all these classes.

3. EXAMPLES OF SELF-DUAL AND ANTI-SELF-DUAL HOMOGENEOUS STRUCTURES

The example we discuss in this section is a special *generalized symmetric space*, more specifically it is a *3-symmetric space*. We recall briefly some basic facts about such spaces and we refer to [3],[4] for proofs and more details.

DEFINITION 3.1. A *family of local cubic diffeomorphisms* on a C^∞ manifold M is a differentiable function $m \mapsto \theta_m$ which assigns to each $m \in M$ a diffeomorphism θ_m on a neighbourhood U(m) of m such that

i) $\theta_m^3 = 1$,

ii) m is the unique fixed point of θ_m.

Next, let θ_{m*} denote the differential of θ_m at m. Then we have

THEOREM 3.2. *Let* M *be a* C^∞ *manifold and assume* $m \mapsto \theta_m$ *is a family of local cubic diffeomorphisms on* M. *Then*

$$\theta_{m*} = -\frac{1}{2} I_m + \frac{\sqrt{3}}{2} J_m \,, \quad m \in M \,, \tag{3.1}$$

defines a C^∞ *almost complex structure* J *on* M

DEFINITION 3.3. The almost complex structure determined by (3.1) is called the *canonical almost complex structure.*

DEFINITION 3.4. A *Riemannian locally 3-symmetric* space (M,g) is a C^∞ Riemannian manifold (M,g) together with a family of local cubic diffeomorphism $m \mapsto \theta_m$ such that each θ_m is a holomorphic isometry in a neighbourhood of p with respect to the canonical almost complex structure of the family.

(M,g) is said to be a *3-symmetric space* when it is connected, locally 3-symmetric and when the domain of definition of each local cubic isometry is all of M.

We have

THEOREM 3.5. *A complete, connected and simply connected Riemannian locally 3-symmetric space is a Riemannian 3-symmetric space.*

Let (M,g,J) be an almost Hermitian manifold. Then (M,g,J) is said to be a *quasi-Kähler* manifold if

(3.2)
$$(\nabla_X J)Y + (\nabla_{JX} J)JY = 0$$

for all $X,Y \in \mathfrak{X}(M)$ and (M,g,J) is said to be *nearly Kählerian* if

$$(\nabla_X J)X = 0$$

for all $X \in \mathfrak{X}(M)$. Further let F denote the Kähler form of (M,g,J), i.e. $F(X,Y) = g(JX,Y)$ for $X,Y \in \mathfrak{X}(m)$. Then (M,g,J) is said to be an *almost Kähler* manifold if F is closed. We have

THEOREM 3.6. *Let* (M,g) *be a locally 3-symmetric space with canonical almost complex structure* J. *Then* (M,g,J) *is a quasi-Kähler manifold.*

The following theorem is well-known and easy to prove.

THEOREM 3.7. *Let* (M,g,J) *be a four-dimensional quasi-Kähler manifold. Then* (M,g,J) *is almost Kählerian.*

Now we return to the homogeneous structures and we consider a 3-symmetric space. The following theorem is proved in [6].

THEOREM 3.8. *Let* (M,g,J) *be a 3-symmetric manifold with canonical almost complex structure* J. *Then the tensor field* T *determined by*

$$T_X Y = \frac{1}{2} J(\nabla_X J)Y \ , \quad X,Y \in \mathfrak{X}(M) \ , \tag{3.3}$$

is a homogeneous structure on (M,g) *and* T *is of type* $\mathcal{T}_2 \oplus \mathcal{T}_3$. *Moreover,* T *is of type* $\mathcal{T}_3$ *(i.e.* (M,g) *is naturally reductive) if and only if* (M,g,J) *is nearly Kählerian and* T *is of type* $\mathcal{T}_2$ *if and only if* (M,g,J) *is almost Kählerian.*

In view of these theorems we consider now four-dimensional simply connected 3-symmetric spaces. These spaces (M,g) are of the following type : (M,g) is the space $\mathbb{R}^4(x,y,u,v)$ with the following metric :

$$g = \{-x + (x^2 + y^2 + 1)^{1/2}\}du^2 + \{x + (x^2 + y^2 + 1)^{1/2}\}dv^2$$

$$- 2ydudv + \lambda^2(1 + x^2 + y^2)^{-1}\{(1 + y^2)dx^2 + (1 + x^2)dy^2 - 2xydxdy\}$$

where λ is a positive constant. The typical symmetry of order 3 at the point $(0,0,0,0)$ is the transformation

$$u' = u \cos\frac{2\pi}{3} - v \sin\frac{2\pi}{3} \ , \quad v' = u \sin\frac{2\pi}{3} + v \cos\frac{2\pi}{3} \ ,$$

$$x' = x \cos\frac{4\pi}{3} - y \sin\frac{4\pi}{3} \ , \quad y' = x \sin\frac{4\pi}{3} + y \cos\frac{4\pi}{3} \ .$$

These spaces are almost Kählerian but they are not nearly Kählerian since (M,g) is not Kählerian with respect to the canonical almost complex structure. Hence the homogeneous structure T given by (3.3) is of type $\mathcal{T}_2$. Next, consider the canonical orientation determined by J. Using a basis (e_1,e_2,Je_1,Je_2) at each point $m \in M$, it follows easily from (3.2) that

$$p_2^+(T) = 0 \ .$$

Hence we have

THEOREM 3.9. *Let* (M,g) *be a four-dimensional simply connected 3-symmetric manifold with the orientation determined by the canonical almost complex structure on* M. *Then the homogeneous structure* T *given by* (3.3) *is anti-self-dual.*

Note that a change of orientation provides a self-dual homogeneous structure.

REFERENCES

[1] Ambrose, W. & Singer, I.M., On homogeneous Riemannian manifolds, *Duke Math. J.* 25 (1958), 647-669.

[2] Atiyah, M., Hitchin, N. & Singer, I.M., Self-duality in four-dimensional Riemannian geometry, *Proc. Roy. Soc. London* A362 (1978), 425-461.

[3] Gray, A., Riemannian manifolds with geodesic symmetries of order 3, *J. Differential Geometry* 7 (1972), 343-369.

[4] Kowalski, O., *Generalized symmetric spaces*, Lecture Notes in Mathematics, 805, Springer-Verlag, Berlin, Heidelberg, New York, 1980.

[5] Singer, I.M. & Thorpe, J.A., The curvature of 4-dimensional Einstein spaces, in *Global Analysis*, Papers in Honor of K. Kodaira, eds. D.C. Spencer & S. Iyanaga, Princeton University Press and University of Tokyo Press, Princeton, 1969, 355-365.

[6] Tricerri, F. & Vanhecke, L., *Homogeneous structures on Riemannian manifolds*, to appear in Lecture Note Series, London Math. Soc., 1983.

[7] Weyl, H., *Classical groups, their invariants and representations*, Princeton University Press, Princeton, 1946.

Politecnico di Torino
Dipartimento di Matematica
Corso Duca degli Abruzzi 24
10129 Torino, Italia

Katholieke Universiteit Leuven
Departement Wiskunde
Celestijnenlaan 200 B
B-3030 Leuven, Belgium

/ol. 900: P. Deligne, J. S. Milne, A. Ogus, and K.-Y. Shih, Hodge Cycles, Motives, and Shimura Varieties. V, 414 pages. 1982.

Vol. 901: Séminaire Bourbaki vol. 1980/81 Exposés 561–578. III, 299 pages. 1981.

/ol. 902: F. Dumortier, P.R. Rodrigues, and R. Roussarie, Germs of Diffeomorphisms in the Plane. IV, 197 pages. 1981.

Vol. 903: Representations of Algebras. Proceedings, 1980. Edited by M. Auslander and E. Lluis. XV, 371 pages. 1981.

/ol. 904: K. Donner, Extension of Positive Operators and Korovkin Theorems. XII, 182 pages. 1982.

/ol. 905: Differential Geometric Methods in Mathematical Physics. Proceedings, 1980. Edited by H.-D. Doebner, S.J. Andersson, and H.R. Petry. VI, 309 pages. 1982.

/ol. 906: Séminaire de Théorie du Potentiel, Paris, No. 6. Proceedings. Edité par F. Hirsch et G. Mokobodzki. IV, 328 pages. 1982.

/ol. 907: P. Schenzel, Dualisierende Komplexe in der lokalen Algebra und Buchsbaum-Ringe. VII, 161 Seiten. 1982.

/ol. 908: Harmonic Analysis. Proceedings, 1981. Edited by F. Ricci and G. Weiss. V, 325 pages. 1982.

/ol. 909: Numerical Analysis. Proceedings, 1981. Edited by J.P. Hennart. VII, 247 pages. 1982.

/ol. 910: S.S. Abhyankar, Weighted Expansions for Canonical Desingularization. VII, 236 pages. 1982.

/ol. 911: O.G. Jørsboe, L. Mejlbro, The Carleson-Hunt Theorem on Fourier Series. IV, 123 pages. 1982.

/ol. 912: Numerical Analysis. Proceedings, 1981. Edited by G. A. Watson. XIII, 245 pages. 1982.

ol. 913: O. Tammi, Extremum Problems for Bounded Univalent Functions II. VI, 168 pages. 1982.

ol. 914: M. L. Warshauer, The Witt Group of Degree k Maps and Asymmetric Inner Product Spaces. IV, 269 pages. 1982.

ol. 915: Categorical Aspects of Topology and Analysis. Proceedings, 1981. Edited by B. Banaschewski. XI, 385 pages. 1982.

ol. 916: K.-U. Grusa, Zweidimensionale, interpolierende Lg-Splines und ihre Anwendungen. VIII, 238 Seiten. 1982.

ol. 917: Brauer Groups in Ring Theory and Algebraic Geometry. Proceedings, 1981. Edited by F. van Oystaeyen and A. Verschoren. VIII, 00 pages. 1982.

ol. 918: Z. Semadeni, Schauder Bases in Banach Spaces of Continuous Functions. V, 136 pages. 1982.

ol. 919: Séminaire Pierre Lelong – Henri Skoda (Analyse) Années 1980/81 et Colloque de Wimereux, Mai 1981. Proceedings. Edité par P. Lelong et H. Skoda. VII, 383 pages. 1982.

ol. 920: Séminaire de Probabilités XVI, 1980/81. Proceedings. Edité par J. Azéma et M. Yor. V, 622 pages. 1982.

ol. 921: Séminaire de Probabilités XVI, 1980/81. Supplément: Géométrie Différentielle Stochastique. Proceedings. Edité par J. Azéma M. Yor. III, 285 pages. 1982.

ol. 922: B. Dacorogna, Weak Continuity and Weak Lower Semicontinuity of Non-Linear Functionals. V, 120 pages. 1982.

ol. 923: Functional Analysis in Markov Processes. Proceedings, 81. Edited by M. Fukushima. V, 307 pages. 1982.

ol. 924: Séminaire d'Algèbre Paul Dubreil et Marie-Paule Malliavin. Proceedings, 1981. Edité par M.-P. Malliavin. V, 461 pages. 1982.

ol. 925: The Riemann Problem, Complete Integrability and Arithmetic Applications. Proceedings, 1979-1980. Edited by D. Chudnovsky and G. Chudnovsky. VI, 373 pages. 1982.

ol. 926: Geometric Techniques in Gauge Theories. Proceedings, 81. Edited by R. Martini and E.M.de Jager. IX, 219 pages. 1982.

Vol. 927: Y. Z. Flicker, The Trace Formula and Base Change for GL (3). XII, 204 pages. 1982.

Vol. 928: Probability Measures on Groups. Proceedings 1981. Edited by H. Heyer. X, 477 pages. 1982.

Vol. 929: Ecole d'Eté de Probabilités de Saint-Flour X – 1980. Proceedings, 1980. Edited by P.L. Hennequin. X, 313 pages. 1982.

Vol. 930: P. Berthelot, L. Breen, et W. Messing, Théorie de Dieudonné Cristalline II. XI, 261 pages. 1982.

Vol. 931: D.M. Arnold, Finite Rank Torsion Free Abelian Groups and Rings. VII, 191 pages. 1982.

Vol. 932: Analytic Theory of Continued Fractions. Proceedings, 1981. Edited by W.B. Jones, W.J. Thron, and H. Waadeland. VI, 240 pages. 1982.

Vol. 933: Lie Algebras and Related Topics. Proceedings, 1981. Edited by D. Winter. VI, 236 pages. 1982.

Vol. 934: M. Sakai, Quadrature Domains. IV, 133 pages. 1982.

Vol. 935: R. Sot, Simple Morphisms in Algebraic Geometry. IV, 146 pages. 1982.

Vol. 936: S.M. Khaleelulla, Counterexamples in Topological Vector Spaces. XXI, 179 pages. 1982.

Vol. 937: E. Combet, Intégrales Exponentielles. VIII, 114 pages. 1982.

Vol. 938: Number Theory. Proceedings, 1981. Edited by K. Alladi. IX, 177 pages. 1982.

Vol. 939: Martingale Theory in Harmonic Analysis and Banach Spaces. Proceedings, 1981. Edited by J.-A. Chao and W.A. Woyczyński. VIII, 225 pages. 1982.

Vol. 940: S. Shelah, Proper Forcing. XXIX, 496 pages. 1982.

Vol. 941: A. Legrand, Homotopie des Espaces de Sections. VII, 132 pages. 1982.

Vol. 942: Theory and Applications of Singular Perturbations. Proceedings, 1981. Edited by W. Eckhaus and E.M. de Jager. V, 363 pages. 1982.

Vol. 943: V. Ancona, G. Tomassini, Modifications Analytiques. IV, 120 pages. 1982.

Vol. 944: Representations of Algebras. Workshop Proceedings, 1980. Edited by M. Auslander and E. Lluis. V, 258 pages. 1982.

Vol. 945: Measure Theory. Oberwolfach 1981, Proceedings. Edited by D. Kölzow and D. Maharam-Stone. XV, 431 pages. 1982.

Vol. 946: N. Spaltenstein, Classes Unipotentes et Sous-groupes de Borel. IX, 259 pages. 1982.

Vol. 947: Algebraic Threefolds. Proceedings, 1981. Edited by A. Conte. VII, 315 pages. 1982.

Vol. 948: Functional Analysis. Proceedings, 1981. Edited by D. Butković, H. Kraljević, and S. Kurepa. X, 239 pages. 1982.

Vol. 949: Harmonic Maps. Proceedings, 1980. Edited by R.J. Knill, M. Kalka and H.C.J. Sealey. V, 158 pages. 1982.

Vol. 950: Complex Analysis. Proceedings, 1980. Edited by J. Eells. IV, 428 pages. 1982.

Vol. 951: Advances in Non-Commutative Ring Theory. Proceedings, 1981. Edited by P.J. Fleury. V, 142 pages. 1982.

Vol. 952: Combinatorial Mathematics IX. Proceedings, 1981. Edited by E. Billington, S. Oates-Williams, and A.P. Street. XI, 443 pages. 1982.

Vol. 953: Iterative Solution of Nonlinear Systems of Equations. Proceedings, 1982. Edited by R. Ansorge, Th. Meis, and W. Törnig. VII, 202 pages. 1982.

Vol. 954: S.G. Pandit, S.G. Deo, Differential Systems Involving Impulses. VII, 102 pages. 1982.

Vol. 955: G. Gierz, Bundles of Topological Vector Spaces and Their Duality. IV, 296 pages. 1982.

Vol. 956: Group Actions and Vector Fields. Proceedings, 1981. Edited by J.B. Carrell. V, 144 pages. 1982.

Vol. 957: Differential Equations. Proceedings, 1981. Edited by D.G. de Figueiredo. VIII, 301 pages. 1982.

Vol. 958: F.R. Beyl, J. Tappe, Group Extensions, Representations, and the Schur Multiplicator. IV, 278 pages. 1982.

Vol. 959: Géométrie Algébrique Réelle et Formes Quadratiques, Proceedings, 1981. Edité par J.-L. Colliot-Thélène, M. Coste, L. Mahé, et M.-F. Roy. X, 458 pages. 1982.

Vol. 960: Multigrid Methods. Proceedings, 1981. Edited by W. Hackbusch and U. Trottenberg. VII, 652 pages. 1982.

Vol. 961: Algebraic Geometry. Proceedings, 1981. Edited by J.M. Aroca, R. Buchweitz, M. Giusti, and M. Merle. X, 500 pages. 1982.

Vol. 962: Category Theory. Proceedings, 1981. Edited by K.H. Kamps, D. Pumplün, and W. Tholen, XV, 322 pages. 1982.

Vol. 963: R. Nottrot, Optimal Processes on Manifolds. VI, 124 pages. 1982.

Vol. 964: Ordinary and Partial Differential Equations. Proceedings, 1982. Edited by W.N. Everitt and B.D. Sleeman. XVIII, 726 pages. 1982.

Vol. 965: Topics in Numerical Analysis. Proceedings, 1981. Edited by P.R. Turner. IX, 202 pages. 1982.

Vol. 966: Algebraic K-Theory. Proceedings, 1980, Part I. Edited by R.K. Dennis. VIII, 407 pages. 1982.

Vol. 967: Algebraic K-Theory. Proceedings, 1980. Part II. VIII, 409 pages. 1982.

Vol. 968: Numerical Integration of Differential Equations and Large Linear Systems. Proceedings, 1980. Edited by J. Hinze. VI, 412 pages. 1982.

Vol. 969: Combinatorial Theory. Proceedings, 1982. Edited by D. Jungnickel and K. Vedder. V, 326 pages. 1982.

Vol. 970: Twistor Geometry and Non-Linear Systems. Proceedings, 1980. Edited by H.-D. Doebner and T.D. Palev. V, 216 pages. 1982.

Vol. 971: Kleinian Groups and Related Topics. Proceedings, 1981. Edited by D.M. Gallo and R.M. Porter. V, 117 pages. 1983.

Vol. 972: Nonlinear Filtering and Stochastic Control. Proceedings, 1981. Edited by S.K. Mitter and A. Moro. VIII, 297 pages. 1983.

Vol. 973: Matrix Pencils. Proceedings, 1982. Edited by B. Kågström and A. Ruhe. XI, 293 pages. 1983.

Vol. 974: A. Draux, Polynômes Orthogonaux Formels – Applications. VI, 625 pages. 1983.

Vol. 975: Radical Banach Algebras and Automatic Continuity. Proceedings, 1981. Edited by J.M. Bachar, W.G. Bade, P.C. Curtis Jr., H.G. Dales and M.P. Thomas. VIII, 470 pages. 1983.

Vol. 976: X. Fernique, P.W. Millar, D.W. Stroock, M. Weber, Ecole d'Eté de Probabilités de Saint-Flour XI – 1981. Edited by P.L. Hennequin. XI, 465 pages. 1983.

Vol. 977: T. Parthasarathy, On Global Univalence Theorems. VIII, 106 pages. 1983.

Vol. 978: J. Ławrynowicz, J. Krzyż, Quasiconformal Mappings in the Plane. VI, 177 pages. 1983.

Vol. 979: Mathematical Theories of Optimization. Proceedings, 1981. Edited by J.P. Cecconi and T. Zolezzi. V, 268 pages. 1983.

Vol. 980: L. Breen. Fonctions thêta et théorème du cube. XIII, 115 pages. 1983.

Vol. 981: Value Distribution Theory. Proceedings, 1981. Edited by I. Laine and S. Rickman. VIII, 245 pages. 1983.

Vol. 982: Stability Problems for Stochastic Models. Proceedings, 1982. Edited by V.V. Kalashnikov and V.M. Zolotarev. XVII, 295 pages. 1983.

Vol. 983: Nonstandard Analysis-Recent Developments. Edited A.E. Hurd. V, 213 pages. 1983.

Vol. 984: A. Bove, J.E. Lewis, C. Parenti, Propagation of Singulariti for Fuchsian Operators. IV, 161 pages. 1983.

Vol. 985: Asymptotic Analysis II. Edited by F. Verhulst. VI, 497 pag 1983.

Vol. 986: Séminaire de Probabilités XVII 1981/82. Proceedin Edited by J. Azéma and M. Yor. V, 512 pages. 1983.

Vol. 987: C.J. Bushnell, A. Fröhlich, Gauss Sums and p-adic Divisi Algebras. XI, 187 pages. 1983.

Vol. 988: J. Schwermer, Kohomologie arithmetisch definierter Gru pen und Eisensteinreihen. III, 170 pages. 1983.

Vol. 989: A.B. Mingarelli, Volterra-Stieltjes Integral Equations a Generalized Ordinary Differential Expressions. XIV, 318 pages. 198

Vol. 990: Probability in Banach Spaces IV. Proceedings, 198 Edited by A. Beck and K. Jacobs. V, 234 pages. 1983.

Vol. 991: Banach Space Theory and its Applications. Proceedin 1981. Edited by A. Pietsch, N. Popa and I. Singer. X, 302 pag 1983.

Vol. 992: Harmonic Analysis, Proceedings, 1982. Edited by G. Ma ceri, F. Ricci and G. Weiss. X, 449 pages. 1983.

Vol. 993: R.D. Bourgin, Geometric Aspects of Convex Sets with t Radon-Nikodým Property. XII, 474 pages. 1983.

Vol. 994: J.-L. Journé, Calderón-Zygmund Operators, Pseudo-D ferential Operators and the Cauchy Integral of Calderón. VI, 1 pages. 1983.

Vol. 995: Banach Spaces, Harmonic Analysis, and Probability Theo Proceedings, 1980–1981. Edited by R.C. Blei and S.J. Sidn V, 173 pages. 1983.

Vol. 996: Invariant Theory. Proceedings, 1982. Edited by F. Ghera delli. V, 159 pages. 1983.

Vol. 997: Algebraic Geometry – Open Problems. Edited by C. C berto, F. Ghione and F. Orecchia. VIII, 411 pages. 1983.

Vol. 998: Recent Developments in the Algebraic, Analytical, a Topological Theory of Semigroups. Proceedings, 1981. Edited K.H. Hofmann, H. Jürgensen and H.J. Weinert. VI, 486 pages. 198

Vol. 999: C. Preston, Iterates of Maps on an Interval. VII, 205 page 1983.

Vol. 1000: H. Hopf, Differential Geometry in the Large, VII, 184 page 1983.

Vol. 1001: D.A. Hejhal, The Selberg Trace Formula for PSL(2, I Volume 2. VIII, 806 pages. 1983.

Vol. 1002: A. Edrei, E.B. Saff, R.S. Varga, Zeros of Sections Power Series. VIII, 115 pages. 1983.

Vol. 1003: J. Schmets, Spaces of Vector-Valued Continuous Fun tions. VI, 117 pages. 1983.

Vol. 1004: Universal Algebra and Lattice Theory. Proceedings, 198 Edited by R.S. Freese and O.C. Garcia. VI, 308 pages. 1983.

Vol. 1005: Numerical Methods. Proceedings, 1982. Edited by V. P reyra and A. Reinoza. V, 296 pages. 1983.

Vol. 1006: Abelian Group Theory. Proceedings, 1982/83. Edited b R. Göbel, L. Lady and A. Mader. XVI, 771 pages. 1983.

Vol. 1007: Geometric Dynamics. Proceedings, 1981. Edited by J. Pal Jr. IX, 827 pages. 1983.

Vol. 1008: Algebraic Geometry. Proceedings, 1981. Edited by J. Do gachev. V, 138 pages. 1983.

Vol. 1009: T.A. Chapman, Controlled Simple Homotopy Theory an Applications. III, 94 pages. 1983.

Vol. 1010: J.-E. Dies, Chaînes de Markov sur les permutations. I 226 pages. 1983.